Impeaching Mere Creationism

Philip Frymire

Writers Club Press

San Jose New York Lincoln Shanghai

Impeaching Mere Creationism

Published by Writers Club Press
an imprint of iUniverse.com, Inc.

For information address:
iUniverse.com, Inc.
620 North 48th Street
Suite 201
Lincoln, NE 68504-3467
www.iuniverse.com

ISBN: 0-595-00196-3

Printed in the United States of America

To students, may you take biology classes that are intelligently designed.

Contents

Introduction

This book was brought to fruition by several factors. The first was irritation provided by reading Phillip E. Johnson's *An Easy-to-Understand Guide for Defeating Darwinism by Opening Minds* a couple of years ago. Johnson seems to be the leader of the recent creationist resurgence. The arguments Johnson makes in the book are ludicrous to anyone with any training in evolutionary biology, but the book is aimed at high school students and others with little or no background in science or critical thinking. This irritation festered for two years. Then the Kansas State Board of Education approved its by now infamous standards which deleted almost all references to evolution. A number of quotes from the creationists involved in this action had a distinctively Johnsonian ring to them. Then, in November of 1999, in my home state of Oklahoma, the Oklahoma State Textbook Committee voted to require that a ridiculous evolution disclaimer be pasted into all biology textbooks approved for use in the state. (Subsequently, the state attorney general ruled that the committee had no authority to require such a disclaimer and that it had violated the state's "open meeting act" by not bothering to include the disclaimer on its posted agenda. Legislation has since been introduced which would give the committee authority to require a disclaimer. In the meantime, the committee has rejected five textbooks as being unacceptable for use in the state. Stay tuned. The drama is continuing.) This led to a predictable spate of anti-evolution twaddle in the "letters to the editor" section of the local newspaper (the *Tulsa World*). Most of these letters betrayed a truly phenomenal level of ignorance concerning evolutionary biology. It never ceases to amaze me how everyone feels qualified to pontificate on evolution. You never see letters criticizing quantum physics or atomic theory or the germ theory of disease. But everyone can fill you in on the "problems" with evolution.

I am a petroleum geologist by profession, but I have a degree in zoology as well as geology. I was fascinated by evolutionary biology and animal behavior while I was in college and that interest has been with me ever since. So while I am not a practicing professional zoologist, I still enjoy keeping up with the latest developments in evolutionary theory. I agree wholeheartedly with Richard Dawkins's assessment:

> Knowledge of evolution may not be strictly useful in everyday commerce. You can live some sort of life and die without ever hearing the name of Darwin. But if, before you die, you want to understand why you lived in the first place, Darwinism is the one subject that you must study.

And with George C. Williams:

> Natural selection is a process of pervasive importance in the biological world, which includes our own species, and on which that species is utterly dependent. Progress in evolutionary biology and its applications is perhaps most obviously relevant to medical and environmental issues, but there is no aspect of human life for which an understanding of evolution is not a vital necessity.

So I was somewhat bemused to see Johnson's book prominently displayed on the new books table at a local bookstore a couple of years ago. On the back cover was effusive praise, including "Phillip Johnson is our age's clearest thinker on the issue of evolution and its impact on society." I had never read a creationist book, although I had read a few books written by professional scientists replying to creationist arguments. I decided to give *Defeating Darwinism* a shot. After reading the book and finding nothing useful concerning evolution I read Johnson's *Darwin on Trial* and *Reason in the Balance* since these were supposedly more "technical" and I might have missed something in the "easy-to-understand guide." I couldn't find anything new or substantive concerning evolution in these books either. As Martin Gardner has noted, Johnson's objections to Darwinism are "moth-eaten."

Subsequently, I watched a debate on William F. Buckley's *Firing Line* between creationists (including Johnson, Michael Behe, David Berlinski, and Buckley) and evolutionists (including Kenneth Miller, Barry Lynn, Eugenie Scott, and Michael Ruse). Creationists love the debate format. The audience is often ignorant of the evidence at best and openly hostile to evolution at worst. Creationists can attack evolution and score debating points and, most importantly, not give any specific alternative to evolution. Alas, this didn't work in court cases involving the creation/evolution controversy. There you had to put on a positive scientific case. There the creationists were soundly defeated. I thought the debate itself was pretty pointless. The creationists mostly railed against "naturalism" and "materialism" and bashed Richard Dawkins, who wasn't even there. Kenneth Miller made some excellent points, but the evolution side seemed most concerned with avoiding the label of "materialists," presumably to avoid offending religious sensibilities. To their credit, the evolutionists attempted to get the creationists to give their version of natural history. Predictably, the creationists offered nothing. I was surprised that people thought the debate was useful. Evidence, not debates, ultimately decides scientific issues. If people are concerned about this issue they need to get up and go investigate the publicly available evidence themselves.

After reading these books and watching the debate I was amazed that anyone would take this stuff ("Mere Creation," as Johnson and others call it) as a serious alternative to modern evolutionary theory. Professional scientists, of course, don't. They have written several book length responses to creationism, but on the whole they seem to follow a policy of ignoring it, presumably to avoid giving the creationists what they want the most, publicity. Still, there are many detailed refutations of creationism available on the internet and elsewhere (see the notes for a listing of a few). Robert Pennock's *Tower of Babel: The Evidence Against the New Creationism* came out in early 1999. It is an excellent and comprehensive refutation of recent (and some not so recent) creationist claims. Kenneth Miller's *Finding Darwin's God* came out in the fall of 1999. It too is a devastating critique of Johnson's views as well as those of other creationists. I have no illusions of improving on the detailed scholarly refutations already out there. However, I thought it might be useful to write a short, less technical, common sense (and hopefully easily understood) rebuttal to Johnson's claims aimed at the same audience he targeted: high school and beginning college students as well as their parents and teachers.

This comes to the crux of the matter. I would hope that all high school science students would be able to experience the fascination of evolutionary biology. The National Center for Science Education does a lot of good work in this area. I know my exposure to evolution in high school biology was perfunctory at best. Students need to be spared the mind-numbing instruction in biology which all too often consists of mostly memorization with no theoretical framework with which to make sense of what they are learning.

Unfortunately, students aren't the only ones who could use an education in evolution. I am constantly surprised by the number of seemingly intelligent, educated people who are utterly clueless when it comes to evolution. No doubt they simply haven't been exposed to it.

I will not in these pages give any detailed recitation of the abundant publicly available evidence for evolution and natural selection. There are many good sources available (see the notes for a few). I am primarily interested in responding to Johnson's attacks. Still, some brief definitions are in order. When I refer to evolution I mean "descent with modification" (that all living things we see today have a common ancestor to which they are connected by an unbroken chain of modified ancestors).

Natural selection is Darwin's mechanism for generating (and explaining) adaptations (features like eyes, ears, etc. that function as aids to survival and reproduction). There is a "struggle for existence" in the natural world due to excess production of offspring and limited resources. Living organisms exhibit heritable variation in individual characteristics. Genetic mutation is the ultimate source of variation. Sexual recombination magnifies variation. At least some of this variation will result in certain organisms being better suited to their environment, and so more likely to survive and reproduce. This is natural selection. Mutations are random with respect to whether they are beneficial to the organism or not. Natural selection is definitely nonrandom. It is also cumulative. It cumulatively preserves what has "worked" in ancestral environments. Natural selection is not the only means of evolution and not all features of organisms are adaptations. Genetic drift (random sampling effects on gene frequencies), selectively neutral molecular change and historical factors are also involved. Natural selection *is*, however, the only known mechanism for generating adaptations.

I am not attacking anyone's religious beliefs. What I am attacking is Johnson's anti-evolution arguments and his insistence that we need to use supernatural explanations in science. These arguments are being used to subvert science education. My purpose is to show that his arguments are ludicrous when examined with just basic common sense. In separate chapters, we will examine his views on naturalism, macroevolution, fossils, intelligent design, human evolution, his avoidance of the real world, and the supposed evil effects of evolution on society. A summary chapter near the end of the book will serve to illuminate his overall strategy and where I believe he's coming from. My goals in writing the book are to show why Johnson's (and other creationists') arguments are silly, but more importantly, I will be delighted if I can encourage readers to go out and learn about evolutionary biology themselves. As I mentioned earlier, if people are concerned about the creation/evolution controversy, they have no excuse for not researching the evidence for themselves.

Finally, I must confess to one other provocation that led me to want to write this book. Johnson believes that being a lawyer gives him special insight in analyzing the logic of arguments. He believes that evolutionary biology is full of bad reasoning and that the whole field is illogical. In fact, it is his status as a law professor at a respected secular university that has made him the leader of the "new creationism." I have some experience as an "expert" witness in protested cases at the Oklahoma Corporation Commission. (Oklahoma and other states have various regulations concerning oil and gas well spacing, distances from lease lines, etc. In some cases companies must file for relief when they want to drill additional wells in a spacing unit or drill closer to lease lines than regulations allow. Occasionally other parties object to these filings and the result is a protested hearing complete with attorneys, geological witnesses and exhibits.) I have been cross-examined (and seen other witnesses cross-examined) many times by lawyers. I believe I have some insight in analyzing the tactics used by lawyers when they have a weak case.

The Materialist Conspiracy

Johnson's main claim in his attack on evolution (and all of science) is that the evidence in support of both descent with modification and natural selection is weak and that the only reason that biologists accept evolution is that they have a "naturalistic" or "materialistic" bias. He has compared the National Academy of Sciences to the College of Cardinals which harassed Galileo. According to Johnson, in the same way that the Church put down Galileo's heresy, Darwinists today immediately attempt to put down anyone who challenges their naturalistic bias.

The attacks on the evidence will be taken up later. For now, what about this evil materialist bias? The usual answer given is to draw a distinction between "methodological naturalism" and "ontological naturalism." Methodological naturalism simply means that science looks for "natural" explanations of phenomena (based on matter, energy, laws of physics and chemistry, etc.). Supernatural elements are out of bounds. Ontological naturalism says that nature is all there is (there is no supernatural). Johnson maintains that this is what biologists are assuming. By doing so, he claims, they are ruling out the supernatural and thereby declaring that there has to be a natural explanation for life by definition. In this way, he says, they are able to simply ignore a lack of evidence. This applies to both descent with modification and natural selection. In his words, "Universal common ancestry is as much a product of materialist philosophy as is the mutation/selection mechanism." In his view, if scientists would only remove their naturalistic blinders they would conclude that we were created by "…a God who acted openly and who left his fingerprints all over the evidence."

Now, science is certainly guilty of methodological naturalism, but I don't know how many scientists there are who flat out deny the possibility of the supernatural. It is certainly possible to be a scientist and still have supernatural beliefs. More interesting to me is why Johnson thinks a supernatural explanation for life is a good explanation. We'll get to his nonexistent "evidence" later. As Richard Dawkins and others have pointed out over and over, the wildly improbable organized complexity of life is the thing that biology attempts to explain. To invoke God (or aliens, unspecified intelligent designers, etc.) doesn't explain anything. The intelligent designer is surely more complicated than his creation. If you can assume God was always there, why not assume that complex life was always there? As a child would ask, "Who made God?" You have simply moved the problem one step back. A supernatural "explanation" is the lazy way out.

Does this prove that the supernatural isn't involved? Of course not. But if you want to use it in science you had better have an excellent *positive* case for its involvement. *Of course* scientists look for naturalistic mechanisms as explanations. Why? Aside from philosophical reasons concerning scientific method which I don't care to get into, the common sense reason is quite simple: no one has ever produced any objective scientific evidence for the involvement of the supernatural. Supernaturalism has been an impediment to science because it explains nothing and suggests no new lines of research. There has never been a case in the whole history of science where invoking the supernatural has led to any scientific advance. Creationism is no exception. Johnson argues against naturalism not only in evolutionary biology but in all fields of science. Consider his views on the mind:

> For one thing, materialism applied to the mind undermines the validity of all reasoning, including one's own. If our theories are

the products of chemical reactions, how can we know whether our theories are true? Perhaps Richard Dawkins believes in Darwinism only because he has a certain chemical in his brain, and his belief could be changed by somehow inserting a different chemical.

Johnson seems to consider this a knockdown argument against materialism. But consider this slightly revised version:

> For one thing, supernaturalism applied to the mind undermines the validity of all reasoning, including one's own. If our theories are the products of souls, spirits, or whatever non-material entity you prefer, how can we know whether our theories are true? Perhaps Phillip Johnson believes in intelligent design only because he has a certain soul in his brain (or wherever you think the soul resides), and his belief could be changed by somehow inserting a different soul.

The way we determine whether our theories are true is by testing them against the real world. Richard Dawkins is a Darwinist because of massive supporting evidence and because natural selection is the only known theory which can in principle explain complex adaptations. As for the mind, I'm sure neurophysiologists will embrace Johnson's theory as soon as he or anyone else supplies even a shred of evidence for it. Possible examples might include people who continue reasoning despite having their oxygen cut off, or being subjected to anesthesia, or having their head crushed. Most people have supernatural beliefs. The sensible ones know that these beliefs do not belong in science.

Should doctors still consider spirits and demons as causes of diseases? Should psychologists invoke spirits and souls in their research? Should auto mechanics consider the supernatural when a car breaks down? Why don't faith healers lay hands on flat tires? In Johnson's own field, the law, should the supernatural be included? Any lawyer who brought up supernatural causes to make a case would be laughed out of court.

These examples seem silly (and they are). The point is that Johnson says that the supernatural is necessary to explain life and that scientists are biased against it. He needs to tell us what his criteria are for when the supernatural is allowed in and when it's not. I think we know what the answer will be.

Naturalism is okay except in those areas of science which happen to yield results which are in conflict with Johnson's particular religious beliefs. He claims to be taking an unbiased look at the evidence, but all of his arguments are negative. As we shall see, he supplies no positive scientific evidence. His only positive arguments consist of quoting Romans and the Gospel of John.

In a strange passage from *Defeating Darwinism*, Johnson criticizes Carl Sagan for supposedly using his "baloney detector" to "…browbeat us into believing, against our better judgment, that humans arose by blind physical and chemical forces over eons from slime." ("Baloney detector" is a phrase that Sagan used in his excellent book, *The Demon Haunted World*, on pseudoscience and other strange beliefs. Sagan basically argued that it is unwise to believe in things like creationism for which there is no evidence or for which there is contrary evidence.) Johnson maintains that if Sagan dropped his naturalistic bias and turned his baloney detector toward evolution he would see that it is untrue. The strange thing about the passage is that it is preceded by an extensive (and justified) praising of physicist Richard Feynman as a fine example for budding scientists. Johnson asks what Feynman would have thought about Sagan's statements concerning evolution. I don't know. But Feynman was a thoroughgoing materialist who accepted evolution and thought there was no particular "purpose" to the universe. In fact, Sagan quoted Feynman at length concerning the "subtlety of matter" in his book (co-written with I.S. Shklovskii) *Intelligent Life in the Universe*. Here is a small excerpt: "When we say we are a pile of atoms, we do not mean we are *merely* a pile of atoms, because a pile of atoms which is not repeated from one to the other might well have the possibilities which you see before you in the mirror (emphasis in the original)." By Johnson's criteria, Feynman would be as deluded and biased as Sagan.

Finally, to see just how silly Johnson's complaints about a "naturalistic bias" are, consider everyday life. In the normal day-to-day world, *everyone* has a "naturalistic bias." I've never met anyone who didn't. I suspect that you haven't either. Even faith healers go to the doctor when they have a medical problem. In everyday life, everyone assumes that there are natural explanations for events. Given that there is no evidence for supernatural involvement in the real, everyday world, which "bias" do you think is more reasonable: natural or supernatural? Johnson himself doesn't practice what he preaches. He chastises scientists for their "naturalistic bias," but in his own profession, the law, he employs a "naturalistic bias." Robert Pennock says it best:

> Thus, despite Johnson's preaching to scientists to incorporate the "possibility" of supernatural interventions in their profession, it is also not very surprising that he never incorporates these in his own professional texts on criminal law. If in some future edition of his textbook on criminal procedure Johnson adds a chapter on how to prosecute witches and lets trial lawyers know how to evaluate "evidence" for the interventions of other supernatural intelligences, then maybe scientists will begin to take him seriously. I expect, however, that we will have to wait for the hexed cows to come home before that day arrives.

A Micromountain from a Macromolehill

Natural selection has been documented over and over in both laboratories and in the field. Rates of evolution have been observed which are orders of magnitude faster than what is necessary to explain changes seen in the fossil record. Indeed, the mystery lies in why so many changes seen in the fossil record are so slow. Johnson, however, insists that natural selection is only capable of causing microevolution (which he defines as "cyclical variation within the type") and not macroevolution (which he defines as "the vaguely described process that supposedly creates innovations such as new complex organs or new body parts"). He repeatedly trashes Jonathan Weiner's book, *The Beak of the Finch*, which won the Pulitzer Prize. This book (which I heartily recommend) contains many lucid descriptions of various scientific studies in which natural selection can be seen operating in a very short time. The centerpiece of the book is the work of Peter and Rosemary Grant on Darwin's finches on the Galapagos Islands. It also includes fascinating examples involving guppies, soapberry bugs, sticklebacks, antibiotic resistance (evolved by bacteria), pesticide resistance (evolved by insects) and many others. One of the defenses evolved by insects against pesticides involves flying off and leaving their legs behind if they happen to land on a leaf covered with pesticide. Weiner includes an amusing quote from scientist Martin Taylor concerning the evolution of pesticide resistance in states whose legislatures are hostile to the teaching of evolution: "These people are trying to ban the teaching of evolution while their own cotton crops are failing because of evolution. How can you be a Creationist farmer any more?"

Johnson, of course, is thoroughly unimpressed by this evidence, dismissing the entire book as being about "finch-beak variation." The book is actually about rapid nonrandom natural selection documented in many different species. Johnson maintains that these examples and all other examples of evolution in action (such examples are routine, despite the surprise from the public which often greets them) are simply microevolution, and have nothing to do with how various plants and animals came to exist. Scientists maintain that these same processes occurring over geologic time have brought about the "macro" changes that Johnson is talking about. (Cumulative selection, does not, of course, explain the entire history of life. Other factors, such as mass extinctions brought about by environmental catastrophes and other random events have a large effect on the history of life. Furthermore, duplications and mutations in important early development genes can have larger effects than changes in other genes. See Carl Zimmer's *At the Water's Edge* for a very readable account of recent developments in macroevolutionary theory.) Johnson thinks this is simply "bad philosophy":

> But what sort of proof is this? If our philosophy demands that small changes add up to big ones, then the scientific evidence is irrelevant. Scientists like to assume that the laws of nature were always and everywhere uniform, because otherwise they could not make inferences about what happened in the distant past or in the opposite end of the universe. They do not assume that the rules which govern activity at one level of magnitude necessarily apply at all other levels. The differences between Newtonian physics, relativity, and quantum mechanics show how unjustified such an assumption would be. What the Darwinists need to supply is not an arbitrary philosophical principle, but a scientific theory of how macroevolution can occur.

But how does he know that macroevolution is at a different "level of magnitude" than microevolution? He simply assumes that it is. It is ridiculous to imply that the differences between Newtonian physics, relativity, and quantum mechanics are analogous to the differences between micro- and macroevolution. None of the evolutionary phenomena in question involves speeds near the speed of light or takes place in the subatomic world. More importantly, remember that the genes in all organisms consist of sequences of "letters" in their DNA (there are four letters: A, T, C, and G, which correspond to the nucleotide bases adenine, thymine, cytosine, and guanine). Change the sequence of letters and you change the organism. Change enough of the letters and you change the organism enough that people would call it a different species. There is no "order of magnitude" difference here. There is a continuum. Any line that you draw separating micro from macro is arbitrary.

It's time to break out some common sense. One concept that is extremely important and helpful in thinking about evolution is almost absurdly simple. Always remember that whether you are dealing with currently living organisms or long dead organisms (fossilized or not) you are dealing with just that: organisms that live, breathe, eat, defecate, copulate, and so on. They all had ancestors (parents, grandparents, great-grandparents, etc.). They may or may not have or have had descendants. There was an unbroken line of real, generation-by-generation animals that lived through all evolutionary transformations, micro or macro. Johnson, of course, would say that this is a naturalistic bias. No, it's a conclusion, supported by a mountain of evidence. "Experiments" further confirming it go on every day. No one has ever observed an animal popping into existence out of thin air. If he has evidence that animals can appear with no ancestors, he needs to show us.

So what does this have to do with micro- vs. macroevolution? Johnson maintains that natural selection cannot change one "type" of animal into another and cannot account for complex organs (more on this later). He thinks that natural selection can only produce small changes within the "type." But as long as animals reproduce there will be mutation and natural selection (and sexual recombination in many species). How much change does there have to be before you go from micro to macro? How many changes does it take? 10? 100? 1000? 1,000,000? He says you can't extrapolate from one to the other. If you can't, why not? Where is his evidence that there is an order of magnitude difference between micro and macro? What specific mechanism will cause genetic mutation to stop? What specific mechanism will cause sexual recombination to stop? What specific mechanism will cause gene duplication and divergence to stop? What specific mechanism will cause natural selection to stop? What specific mechanisms limit change to within the "type?" For that matter, what's a "type?"

Should we apply Johnson's thinking to geology? Geologists can measure tectonic plate motions. They extrapolate these plate motions to explain mid-ocean ridges, mountains, volcanoes, earthquakes, etc. All these phenomena are far more complex than the actual plate motions that have been observed. Should we distinguish between microcontinental drift and macrocontinental drift? Does our naturalistic bias prevent us from seeing the "order of magnitude" difference between them?

What about erosion? Should we distinguish between the "microerosion" we see going on every day and the "macroerosion" that produced the Grand Canyon? What about the "order of magnitude" difference between the "microgrowth" going on at the level of twigs and the mysterious "macrogrowth" which produced the huge boughs near the base of a tree?

I think you can see where this is leading. Johnson will call any evolution observed in the lab or in the wild microevolution. Any evolution not directly observed he will call macroevolution and will be (according to him) mysterious and unexplained. He can then talk about the "missing mechanism" for macroevolution. By using this ploy he makes it impossible to provide any evidence he will admit is convincing. Never mind that this essentially leads to the absurdity of saying that he will not accept macroevolution until someone evolves a human from a single-celled organism in the lab.

You might try bringing up examples of massive adaptive radiation of species in the recent past (well, recent geologically speaking). Weiner mentions fruit flies in Hawaii which have evolved into between 500 and 1,000 different species within the last few million years. These include predators, parasites, nectivores, detritivores, and herbivores. They range in size from pinheads to the length of a child's thumb. He also mentions the cichlid fishes in Lake Victoria which have evolved into about 200 distinct species within 750,000 years. (Incidentally, this radiation of cichlid fishes is the subject of an interesting book by Tijs Goldschmidt: *Darwin's Dreampond*.) Johnson would dismiss both these examples. After all, no one observed them. They are simply inferred from a naturalistic bias. Besides, they're still just fruit flies and fishes.

Picture a Methuselah (maybe through genetic engineering) graduate student of the future who lives for thousands or millions of years. If he doesn't die of boredom, he could observe evolution for far longer than the currently available studies. Would this convince a Johnson of the future? No, it would still be called microevolution.

There is no sense appealing to fossils either because…, but I'm getting ahead of myself. If you are interested in more useful approaches to macroevolution I would recommend the book I mentioned above: *At the Water's Edge*, by Carl Zimmer. He covers two important events in evolution: the first vertebrates to colonize the land, and the return of the ancestors of whales to the water. In fact, *The Beak of the Finch* and *At the Water's Edge* make an excellent tag team in the bout with creationists (although neither was written for that purpose).

The Fossil "Problem"

You may not know it, but evolutionists are committing an egregious sin. They highlight fossils which seem to support evolution, but: "They rarely inform the public about the far greater mass of contrary evidence, such as the absence of ancestors for the major animal groups that appear in the Cambrian explosion." Johnson believes that the fossil record, when viewed without bias, does not support Darwinian evolution. In his words, "It seems for now as if new forms appeared mysteriously and by no known mechanism at various widely separated times in the earth's history." So, Darwinian evolution is out, but, as you can see, Johnson does not tell us what he thinks happened. In fact, Johnson never supplies us with his version of the history of life. There is a good reason for this, and we'll return to this interesting point later, but, for now, what would the fossil record look like if Darwinian evolution were true?

Currently living organisms—with the exception of some single-celled forms—should disappear from the fossil record as you go back in time. That is, if organisms really are the result of descent with modification, then they shouldn't retain the same form throughout geologic history. Furthermore, the further back in time you go, the greater the differences (on average) should be between those fossils and current organisms. Fossils in adjacent strata (layers of rock) should be more similar to each other than fossils from strata which are widely separated in time. Fossils in the youngest rocks should be very similar to living organisms.

There should be a logical progression of forms in the fossil record. The oldest forms should be simple, single-celled aquatic organisms (all life on earth depends on water). Multicellular aquatic forms would follow. Still younger rocks would contain land dwelling organisms. Among vertebrates, fish should appear first, followed by amphibians, reptiles, and mammals (this prediction is based on comparative anatomy). You should not find fossils which are obviously out of sequence. In other words, you should not find fossil organisms in rocks deposited before those organisms could have evolved. Any such fossils would refute Darwinism. It is easy to come up with examples: chimpanzees in the Silurian (a geologic period from about 438 to 408 million years ago), humans in the Precambrian (earlier than about 540 million years ago), and so on. No such fossil has ever been found, although some creationists have made bogus claims that fossil human footprints have been found alongside those of dinosaurs. At least they know what to look for.

The fossil record I have just described is, of course, the fossil record which actually exists. The fossil record of the real world overwhelmingly confirms that evolution has occurred. Johnson maintains that scientists take for granted that evolution is true and then proceed to interpret all the fossil data so that it fits their preconceived bias. This isn't true. To see why, it is important to realize the many different ways the fossil record could have turned out. For instance, it could have turned out that there were no fossil record at all. If species never change, and the earth is very young, there might not be any fossils. Or, if the earth is very old—and species never change—then the fossil record would be a monotonous affair consisting only of currently living species no matter what age rocks you looked in.

There are many other possibilities. Mammals could have shown up first in the Precambrian with bacteria only appearing in very recent rocks. Land organisms could have appeared before aquatic organisms. Reptiles could have appeared before fishes. It could have turned out that there were no extinct forms in the fossil record at all. It could have been that fossils in adjacent strata were not more similar than fossils in widely separated strata. It could have happened that the sequence of appearance of various forms in the record was utterly random.

None of these possible scenarios for how the fossil record could have turned out actually occurred. Instead, the real world fossil record supports descent with modification. Since neither he nor anyone else has ever found an out of sequence fossil, why does Johnson think that the fossil evidence does not support evolution? One of his attacks involves the infamous "Cambrian explosion," the sudden profusion of animal fossils in the Cambrian Period. So far, no one has found any immediate potential ancestors for these diverse forms (see the notes) although there are examples of much older fossils. But does the Cambrian explosion refute evolution? Of course not. The fact that no one has found fossilized potential ancestors for these animals does not mean that they didn't have ancestors. They certainly did, despite Johnson's naturalistic bias complaints. Again, remember our common sense way of thinking about evolution and fossils. Every fossil found in the Cambrian was at one time a living, breathing, eating creature. It had parents and grandparents (and so on) just like every animal alive today. To suggest otherwise contradicts common sense and a mountain of evidence from ongoing everyday life. What Johnson does is exploit confusion surrounding Stephen Jay Gould's book, *Wonderful Life*, to raise an irrelevant smokescreen (again, see the notes).

Do not be deceived by claims that there was something deeply mysterious going on in the Cambrian when supposedly whole new phyla were springing into existence. (Phyla refers to one level in the Linnaean classification system: Kingdom, Phylum, Class, Order, Family, Genus, Species.) Remember that the evolution of life is a branching tree. All splits in the tree of life begin at the species level. It is only after multiple subsequent splits that the arbitrary higher levels in the classification system come into play. Think of the image of a large, old tree. Even the biggest boughs near the base of the trunk began as twigs.

Johnson also claims that mass extinctions documented in the fossil record are a problem for Darwinian evolution. This is just wrong. It is true that Darwin tended to emphasize gradual extinction, but he was at the time continually fighting catastrophism (such as Noah's flood). But Darwinian evolution does not require gradual extinction. Getting blasted by a comet or having your environment devastated by volcanoes does not refute Darwinian evolution although such factors are obviously important in the history of life since they potentially open up many new niches which can then be exploited by surviving organisms. The diversification of mammals after the extinction of the dinosaurs is a classic example. Your ancestors and the ancestors of every organism living today survived generation-by-generation through every mass extinction in the earth's history.

Another confusion exploited by Johnson concerns "punctuated equilibrium." This theory, originally proposed by Niles Eldredge and Stephen Jay Gould in 1972, states that evolution proceeds by short (geologically short, meaning thousands of years) bursts (punctuations) within long periods of stasis (equilibria). In other words, most change is supposed to take place during punctuations with little or no change during the much longer periods of stasis. There is ongoing debate in the scientific community concerning this theory's relative importance. It helps explain the consequences of allopatric speciation (formation of a new species by geographic separation and subsequent genetic divergence of a breeding population) for the fossil record in any particular locality. If a population is split and a small founder group then evolves separately into a new species, the new species may then migrate back into the old area. If this happens, the fossil record will not record the evolution of the new species in the old locality. It will only show its abrupt appearance. Whatever punctuated equilibrium's importance is, it is not non-Darwinian. During a punctuation there is still gradual—although not necessarily constant—change from generation to generation. In other words, there is no evidence that anything mysterious is going on during a punctuation. Darwinian processes still apply. The theory has been jokingly referred to as "evolution by jerks." The counter-joke (for gradual—well, even more gradual—evolution) is "evolution by creeps." Punctuated equilibrium can be thought of as evolution by creepy jerks. The point is that Darwinian evolution doesn't require a constant rate of change although neutral (non-selective) molecular evolution often proceeds that way.

Johnson repeatedly harps on the fact that (as he sees it) there are few, if any, transitional fossils in the geological record. This, of course, is the familiar fossil gaps ploy. There are indeed many gaps in the record but there are also many excellent intermediate forms in the record and more are found all the time. Keep in mind that "transitional" fossils were real organisms that were adapted to the environment of their time. They only become "transitional" in hindsight. Paleontologists of the future will consider organisms living today as "transitional." Also remember that you can never know for certain that an individual fossil was ancestral to another fossil or to a currently living organism. All you can say for sure is that younger fossils (and modern organisms) had ancestors that lived at the same time as the individual fossil in question. An interesting question is: if evolution did not occur (or if it is limited to microevolution within "types"), why are there *any* transitional fossils at all? For that matter, why are there any fossils which differ from currently living forms?

It is extremely difficult to get fossilized in the first place. Think of how few dead animals you have seen lying around. To become fossilized, a dead animal must be buried by sediments before erosion and scavengers can get at the corpse. The geological formation (rock unit) within which the fossil is encased must not be metamorphosed by temperature and pressure or the fossil may be destroyed. (Precambrian rocks are most likely to have been subjected to metamorphosis since they are the oldest.) Even if metamorphosis doesn't get it, a fossil still has to be uplifted and the surrounding rock eroded away before it can even potentially be found. Even then, the fossil is subject to destruction by erosion before anyone notices it. In my own field of petroleum geology, the times of greatest anticipation are when the drilling bit approaches the target zone and when the well logs (various electrical, radioactive, magnetic, sonic, and other measurements made by tools lowered on a wireline into the borehole) are being pulled over the zone of interest. I can imagine the paleontologist's anticipation with every hammer strike, brush stroke, or at seeing a freshly eroded fossiliferous formation after a rainstorm.

One of creationism's favorite gaps has been spectacularly closed within the last few years. It is the evolution of whales from terrestrial mammals. A beautiful series of intermediates has recently been unearthed. I refer you to Carl Zimmer's *At the Water's Edge* and Stephen Jay Gould's essay "Hooking Leviathan by its Past" (in *Dinosaur in a Haystack*) for the fascinating details. This example is particularly amusing since Johnson devotes several paragraphs of *Darwin on Trial* to ridiculing the possibility that whales could have evolved from land mammals. But have no fear. There will always be gaps, and when intermediates are found it only increases the number of gaps (if you had a gap between A and E, then found C, where are B and D?).

Richard Dawkins has an interesting calculation in his book, *Unweaving the Rainbow*. He has us consider our direct female ancestry back to the Cambrian (over 500 million years ago). Assume that every one of our ancestors died atop her mother's grave and was subsequently fossilized. Then assume that every fossil was compressed to a thickness of only one centimeter. How thick a rock sequence must we have to contain our complete, continuous, gapless fossil record? The answer he gets is about 600 miles. That is many times thicker than the entire crust of the earth, and most of the crust's thickness is not sedimentary rock. This calculation gives some perspective on the significance of gaps in the fossil record. It also highlights that "missing link" fans are way to conservative. There are millions of missing links.

So how can we summarize the fossil "problem" in simplest terms? If evolution is true, as you go back in time you should see forms which are different from anything that is living today. You should see some transitional forms. You should not see any fossils grossly out of the sequence we described near the beginning of the chapter. All these criteria are met. Yes, there are gaps. For these, here are your choices:

A) The organisms in the gaps actually lived but they didn't fossilize, were fossilized but later destroyed (either through metamorphosis in the subsurface or erosion on the surface), were fossilized but are still in the subsurface, or were fossilized, are on the surface, but no one has found them yet.

B) Or??? Johnson never tells us his explanation for the gaps. Presumably the organisms on either side of the gaps popped into existence out of thin air. Furthermore, this flashing into existence with no ancestors (now *that's* a missing mechanism) had to have occurred *millions* of times in the past even though, mysteriously, it is never seen to occur in the present.

You make the call.

Intelligent Design

Near the end of the movie *Time Bandits* there is a funny line spoken by Sir Ralph Richardson (playing the Supreme Being). He remarks that it's time to get back to work "...or they'll think I've lost control again and chalk it up to evolution." Johnson maintains that scientists have chalked it up to evolution despite the fact that the evidence, when viewed without naturalistic blinders, shows that the world (and especially humans) has been designed and created by God for a purpose. What that purpose is, Johnson never tells us (more on that later). Other intelligent design advocates are more coy as to the identity of the designer (I am reminded of a Monty Python sketch: "wink, wink, nudge, nudge, say no more, say no more") but the religious element is always there. The argument from design is very old and has been buried many times before. Current intelligent design proponents imply that theirs is a cutting edge science. In fact, it is cutting edge early nineteenth century science (when William Paley's *Natural Theology* was published). Does Johnson have any new arguments to raise it from the ashes?

We have already seen that bringing God into the equation doesn't explain anything. It simply makes the problem worse (where did the intelligent designer come from?). But, of course, it is still possible that a supernatural entity separately created all living things. It's also possible that the universe was created five minutes ago with memories and everything already put in place. What is the evidence that Johnson says so unequivocally indicates intelligent design?

One of his arguments is that living things are just too complex to have evolved their exquisite adaptations (eyes, wings, etc.) by Darwinian natural selection. In other words, Johnson just can't believe that complex things could have evolved by gradual, cumulative selection. Readers of Richard Dawkins will immediately recognize what he has humorously dubbed the "Argument From Personal Incredulity." As Dawkins puts it:

> Never say, and never take seriously anybody who says, "I cannot believe that so-and-so could have evolved by gradual selection."
> I have dubbed this kind of fallacy "the Argument from Personal Incredulity." Time and again, it has proved the prelude to an intellectual banana-skin experience.

(This quote is from *River Out of Eden*. The Argument From Personal Incredulity was originally christened in Dawkins's *The Blind Watchmaker*. Both these books, as well as *Climbing Mount Improbable*, are excellent sources concerning the cumulative power of natural selection to produce incredible adaptive complexity. I encourage you to read Dawkins and see what you think.)

There is a strange pattern to the usage of this argument. You never hear people claiming that intragenomic conflict, tapeworms, malaria parasites, Ebola virus, HIV, tuberculosis, intrauterine cannibalism (in sand tiger sharks, for example), parasites eating their paralyzed host alive from the inside, or any similar phenomena are just too complex to have evolved by natural selection, that they had to have been intelligently designed. Needless to say, the Argument From Personal Incredulity is not one of your great scientific arguments. Nevertheless, it made a big splash with the publication of Michael Behe's *Darwin's Black Box* in 1996.

Behe, unlike Johnson, is a legitimate scientist. However, he is a biochemist, not an evolutionary biologist. Johnson likes to promote the idea that there is a unified intelligent design movement, but there are huge differences between Behe's positions and Johnson's. Johnson not only doesn't accept that natural selection can explain complex adaptations, he does not accept that the history of life consists of descent with modification (that organisms share a common ancestor). Behe finds the idea that organisms have a common ancestor "fairly convincing," but he doesn't appear to accept macroevolution: "Persuasive evidence to support that position, however, has not been forthcoming." This is rather strange since you can't have common ancestry without macroevolution. Behe doesn't take Genesis literally. It's hard to say what Johnson thinks, since he never gives his version of creation. At any rate, if you think there is a unified intelligent design movement that agrees on a particular version of creation, you're mistaken.

Behe maintains that at least some biochemical pathways in living cells are "irreducibly complex" (basically, that if any of the component parts of the pathway is taken out, the entire system collapses). He then concludes that they couldn't have evolved by natural selection. Therefore, they had to be designed. Sound familiar? It's our old friend, the Argument From Personal Incredulity coupled with what philosophers call the "argument from ignorance" (very simply, science hasn't explained this yet, therefore God did it). Behe (and Johnson, who praises Behe's book) is basically making the following argument: I believe this biochemical pathway is irreducibly complex. I cannot believe that it could have evolved in a gradual, Darwinian manner. Therefore, it did not. Therefore, it must be the product of intelligent design. As Johnson says, "Molecular mechanisms, Behe says, are as obviously designed as a spaceship or a computer."

One minor difference springs immediately to mind. We can see who is designing spaceships and computers. We can see them being assembled part by part. No one has ever seen an intelligent designer assembling living organisms. Living organisms aren't assembled from different parts. They grow (develop) from single cells using externally supplied matter and energy in a manner totally different from human constructed artifacts (which probably helps explain why the mechanisms of development have been so difficult for scientists to figure out).

Consider this statement from Johnson: "…in our universal experience unintelligent material processes do not create life…" Think about that for a few seconds. Remember the old joke about being able to mass produce workers using unskilled labor. Our universal experience is that living things are *always* the result of unintelligent material processes. No one has ever observed an intelligent creation of life.

Johnson would probably protest here and say that he is referring to the fact that no living things have ever been observed to arise from non-living things through "unintelligent material processes." That's true, but they have never been observed to arise from "intelligent" processes either. Even if life is originating anywhere on Earth today it would quickly be gobbled up by existing life. As for the original origin of life, there are many theories but, so far, no complete solution. It goes without saying that invoking supernatural causes for the origin of life is just as lazy and useless as invoking them to explain complex adaptations. But the key point is this: evolution means that all living things are the result of a long line of reproduction with modification. Reproduction of life through "unintelligent material processes" is all that has ever been observed. "Intelligent" creation of life that is not the result of a long line of reproduction is something that has never been a part of "our universal experience."

What about the claim that irreducibly complexity cannot evolve by Darwinian selection? The first thing to remember is that Behe's claims are no different from previous "moth-eaten" objections to Darwinism that eyes, wings, etc. could not have evolved through cumulative natural selection. These arguments have been made—and thoroughly answered—since Darwin's time. Using molecules instead of gross anatomy sounds more technical but it is the same argument. Remember that natural selection is a cumulative, nonrandom process. For millions of years random variation has been generated and those variants which had the greatest reproductive success (not just producing offspring, but offspring which produce offspring) have been cumulatively preserved. Natural selection is "blind" in the sense that it has no foresight, but, unlike human designers, it has no biases which limit what it can "try." It can produce incredible complexity, but at the expense of leaving countless inferior "designs" strewn throughout geologic time.

This last point is important. People look at the wonderful adaptations in the natural world and naturally think that there must be some design involved. What they don't see is the far greater number of maladapted or inferior "designs" because those designs are all dead. In every generation only a very select few survive and reproduce. Then another round of variation is served up and selected from. This goes on throughout geologic time, with the number of nonreproductive losers piling up a lot faster than the winners. Individuals do not evolve (despite letters to the editor asking why, if evolution is true, we don't see monkeys turning into men all around us). Gene pools (all the genes in a particular population) evolve. After many generations of cumulative selection, the organisms the genes produce can be radically different from the organisms they produced before the start of the cumulative selection process.

The second thing to notice is that neither Behe nor Johnson has provided any *positive* evidence for intelligent design. They have not discovered any new facts or phenomena. They are simply claiming that certain already known phenomena can't be explained by natural selection.

Lastly, since Behe accepts common ancestry, you might be suspicious of a situation in which evolution by natural selection can seemingly explain the anatomy, physiology, behavior, etc. of organisms, but for some reason supposedly cannot explain the biochemical processes underlying these phenomena.

So, can irreducible complexity evolve by natural selection? Yes, it can. Consider any highly specialized animal, for example, a lion. It has a whole suite of adaptations for an exclusively carnivorous life. It has binocular vision, stalking software, claws, large canines, specialized jaw muscles, and a gut specialized for digesting large quantities of meat. If it lacked any of these traits, it probably couldn't survive. It seems to be an irreducibly complex system. So how can each of these traits evolve step by step until the whole system is in place? The answer is that they don't evolve as individual parts. That is Johnson's and Behe's error. They evolve together. Once a step toward meat eating is taken (whatever it is) then from then on mutations that aid in predation are favored over those for plant eating. As the gut becomes better able to digest meat, longer canines and claws become advantageous in prey capture and killing. These developments encourage selection for binocular vision and stalking software. This in turn fuels selection for further refinements of claws, teeth, and gut. And so on until the complex suite of adaptations is evolved. No doubt the ancestors of lions were once able to get by without having all these exquisite adaptations in their current refined form. Now, however, because of past changes in both the lions themselves and their prey, all parts of the suite of adaptations are necessary, but an observer who arrived to see only the current product might have trouble reconstructing how the entire system evolved.

Obviously, I don't intend this hypothetical example as an exact account of lion evolution. My point is to show that it is a mistake to believe that if you have a system which consists of A, B, C, D, and E, that you must first evolve A, then B, then C, etc. with each component reaching its final form before moving on to the next one. All parts of a system can have variation. It is "utterly fallacious" (as George Williams puts it, getting right to the point, as usual) to argue that complex things cannot evolve by a cumulative series of small changes, each one improving the system slightly. Critics of natural selection's ability to create an eye have often emphasized that small improvements in one part (the retina, for example) would be useless without simultaneous improvements in another part (the lens, for example). But as Williams points out: "An improved retina may be useless without an improved lens, but both retinas and lenses are subject to individual variation. Some of the better retinas would be found in individuals who also had better lenses, so that the improvements, on average, could be favored." The same reasoning applies to any complex organ or system consisting of several interacting parts.

Consider your own heart. It has four chambers and four valves. Take away any of the parts and you'll be in big trouble. It is "irreducibly complex," so it supposedly couldn't have evolved by cumulative natural selection. Yet you can consult any textbook on comparative vertebrate anatomy and you will find (among the hearts of currently living animals) a smooth series of plausible intermediate steps in the evolution of four-chambered hearts.

In fact, irreducible complexity is a routine outcome of evolution. Any vertebrate is full of irreducible complexity. Take out its heart and it's dead. Take out its lungs and it's dead. Yet there is a smooth evolutionary continuum from creatures small enough that simple diffusion can meet their circulatory and respiratory needs all the way to the complicated hearts and lungs in vertebrates. The end product is irreducibly complex, but it can be built by a gradual Darwinian process. I certainly don't know enough biochemistry to evaluate Behe's specific examples, but remember that using molecules instead of gross anatomy doesn't affect the argument about how they could evolve.

Fortunately, of course, there are scientists who do know enough to evaluate Behe's specific claims. They are not impressed. For a devastating critique, see the chapter entitled "God the Mechanic" in Kenneth Miller's *Finding Darwin's God.* One of the things Miller points out is Behe's curious overlooking of comparative biochemistry. One strategy often used in explaining the evolution of eyes, hearts, wings, and other anatomical features is to use comparative anatomy to reveal plausible intermediates in currently living animals. Miller notes that comparing cilia structure and blood clotting in a variety of different animals makes explaining their evolution not nearly as difficult as Behe implies. No doubt the evolution of many biochemical pathways is not completely understood, but surely further research is in order rather than simply giving up, which is exactly what Behe, Johnson, and other design advocates are doing.

I should add that Behe (a scientist, not a lawyer), to his credit, does give a hypothetical scenario for his intelligent design theory:

Suppose that nearly four billion years ago the designer made the first cell, already containing all of the irreducibly complex biochemical systems discussed here and many others. (One can postulate that the designs for systems that were to be used later, such as blood clotting, were present but not "turned on." In present-day organisms plenty of genes are turned off for a while, sometimes for generations, to be turned on at a later time.)

Behe then goes on to explain that the cell with the designed systems could have been left on "autopilot" to "reproduce, mutate, eat and be eaten, bump up against rocks, and suffer all the vagaries of life on earth." Down through the years, chance events might lead to organs becoming nonfunctional or pseudogenes arising. Still, "These chance events do not mean that the initial biochemical systems were not designed. The cellular warts and wrinkles that Miller [referring to Kenneth Miller] takes as evidence of evolution may simply be evidence of age."

Remember that Behe, unlike Johnson, does not deny the obvious overwhelming evidence for the common ancestry of all organisms. Unfortunately, his scenario can't work. As Miller points out, this ancient cell would have had to contain "more than a thousand times the genetic information of one of today's bacteria" if it indeed already contained all the "preformed genetic information required to make a human being" (not to mention all the other organisms which have ever existed). Also, all modern organisms should contain traces of all the genetic information to make any organism, but, mysteriously, they don't. Furthermore, while all those preformed genes are waiting around for hundreds of millions of years before they are expressed, massive amounts of mutations will accumulate since there would be no natural selection to weed harmful mutations out. By the time the genes were needed, they would have been mutated into oblivion. Miller makes the key point:

What this colossal mistake shows is that one cannot develop a narrowly focused anti-evolution argument, like Behe's irreducible complexity, without paying attention to its impact of (*sic*) the other lines of evidence, including the fossil record, that profoundly support evolution. And that evidence, which Behe makes clear he accepts scientifically, squeezes him into a nonsensical position. It forces him, for the sake of consistency, to cobble his acceptance of the earth's well-documented natural history into the idea of God as a biochemical mechanic. The result is an absolutely hopeless genetic fantasy of "preformed" genes waiting for the organisms that might need them to appear gradually—and the utter collapse of Behe's hoped-for biochemical challenge to evolution.

If you haven't figured it out already, you're probably beginning to get an idea why Johnson never gives his version of creation.

A classic example of the step by step evolution of an irreducibly complex system recorded in the fossil record is the evolution of the middle ear bones of mammals from the jaw bones of reptiles. The eardrum is connected to the three small middle ear bones which connect to the inner ear. All parts of the system are necessary for hearing, although you can hear remarkably well with a hole in your eardrum. I've had holes in my right eardrum surgically repaired twice.

If you wish to delve further into irreducible complexity and intelligent design, I would recommend going to "Behe's Empty Box" at http://www.world-of-dawkins.com/box/behe.htm. This is part of an excellent unofficial website on the work of Richard Dawkins maintained by John Catalano. There are many critiques of intelligent design and reviews of Behe's book as well as refutations of his claim that there are no published accounts of the Darwinian evolution of biochemical pathways.

Another intelligent design argument used by Johnson involves the difference between matter and information (between the medium and the message). As he correctly points out, the information in a book is the same whether it is recorded on paper, computer disk, or memorized in someone's head. He then leaps to the conclusion that any complex information (coded in DNA, for example) has to have an author (an intelligent designer in this case).

Now there are many mindless processes that can carry information. Photons carry information about your surroundings as do sound waves. A rock outcrop contains information about the environment in which the rock was deposited. A crime scene is loaded with information. There are no authors involved here. Johnson's objection here boils down to his previous claim that the complexity in living things (in this case, the information in a living thing's DNA) cannot have evolved by cumulative natural selection. We've heard that before.

As part of his intelligent design argument, Johnson particularly attacks two analogies which he says evolutionists use in a deceptive manner. On the contrary, it is Johnson who is deceptive. His first example involves computer programs which generate target phrases through a series of "generations" of selecting from random sequences of letters. In each generation, the sequence closest to the target phrase is selected. Then, randomly varying "offspring" are "bred" from that phrase, selection takes place, and so on, until the target phrase is reached. Johnson claims that: "Richard Dawkins actually uses examples like this to illustrate the creative power of natural selection, and his readers apparently don't see that it's just a trick." Nonsense. Dawkins explicitly states that the example is not an exact analogy to natural selection. Natural selection has no goal in mind. What the example is meant to show is the power of *cumulative* selection as opposed to single step selection. Furthermore, there are now very sophisticated programs that do model natural selection and evolution (yes, the software is intelligently designed, but the organisms in the "cyberecosystem" evolve with no direction or target imposed on them). Thomas Ray, now at my alma mater, the University of Oklahoma, is a leader in this field.

Johnson also attacks analogies made between natural and artificial selection. He claims that such analogies are bogus because artificial selection is intelligent design since someone is doing the selecting. Of course scientists know that in natural selection there is no intelligence involved, and that in artificial selection a selector is involved. The analogy that scientists are making is that many generations of cumulative natural selection can produce tremendous changes in organisms just as many generations of artificial selection have produced prodigious changes in dogs, pigeons, and many other organisms. A selection process that involves only who breeds with whom (or as Carl Sagan and Ann Druyan put it: "meddling with the sex lives of dogs") can and does result in tremendous cumulative change. Johnson especially harps on dog breeds as simply examples of intelligent design. This is ridiculous. No one sat down at a drafting table and designed a dachshund and assembled it from scratch with no ancestors. No one took a wolf and remolded its flesh into a beagle in a single step. I presume that Johnson is trying to insinuate that the only way you could get such prodigious change as you see in the various dog breeds is if you have intelligent design guiding the way. I can picture a creationist of the future jumping for joy at the various fossils in the Quaternary Canid Explosion. He would no doubt place them in different species since in some cases they differ far more than foxes, wolves, and coyotes. There would be this seemingly instantaneous profusion of forms with few or no intermediates to be found.

There is irony in Johnson's insistence that artificial selection is really intelligent design. Some "theistic evolutionists," as best I can tell, believe that God created life by in some way supervising natural selection or controlling the mutations available for natural selection. God would then be the artificial selector and, according to Johnson's interpretation of artificial selection, the intelligent designer. This would seem a nice fit for Johnson. Unfortunately, he can't stand theistic evolutionists.

Now we come to the most ridiculous part of Johnson's supposedly scientific theory of intelligent design. He tells us that we must: "Learn to distinguish between theories that put themselves at risk—that is, invite testing by observation or experiment—and theories that can't be shown to be either true or false." This is excellent advice. We have already seen how an out of place fossil could refute evolution. If the earth were only a few thousand years old, evolution would be out the window (there wouldn't have been enough time). You could refute natural selection by finding an adaptation in an organism which existed solely for the benefit of another organism. One example would be gazelles which approach cheetahs and remain motionless while extending their necks out. Another would be horses with built-in saddles. If you could find any adaptation which was not for the benefit of the genes which produced it, you could refute natural selection. Another way to refute natural selection would be to find complex organisms that are not a part of a long line of reproduction (such as animals that lack reproductive organs or animals that spontaneously assemble themselves or pop into existence from nothing). Find an adaptation that could not possibly have been created by cumulative slight modifications and natural selection will be refuted. Neither Johnson nor any other creationist has ever supplied us with anything like these examples.

Evolution and natural selection can and have been tested. What about intelligent design theory? Consider Johnson's reasoning here:

> If I say I believe in creation on faith, no matter what the evidence is, then we can't test my belief by scientific observations or experiments. But if I say the evidence indicates that living organisms are necessarily the products of intelligent design and that life never could have emerged by purely natural means from a prebiotic soup of chemicals, my statement invites scientific testing. Theories of chemical and biological evolution aim to contradict my hypothesis of intelligent design, by showing that purposeless natural processes can do the creating by evolution. The question is whether they have been successful in doing this—that is, whether the theories have passed the experimental test or failed it.

Johnson has slickly evaded the issue. Intelligent design theory makes absolutely no predictions about living things. Martin Daly and Margo Wilson, in their superb book, *Homicide*, say it best (after first noting that there are many different creation stories): "The more serious problem is that creationism is simply devoid of empirical implications. Whatever turns up must be the will of the creator(s). Implications for the practical investigation of the natural world are nil." We will return to this point later, but for now consider Johnson's reasoning again. Here is a rough translation:

> I believe that life is the result of intelligent design. I have no evidence to back up this claim nor does my theory make any predictions. My theory cannot be tested, but I will make it look like it can by saying that you have to prove me wrong by showing that evolution by natural selection can explain the complexity of life. I will invoke the Argument From Personal Incredulity to show that it can't do the job. The Argument From Ignorance will then prove that my theory is correct.

Here's another way to see how ridiculous Johnson's reasoning is here. Imagine you had a theory which said germ B causes disease B. You work long and hard accumulating evidence that germ B really does cause disease B. You make various predictions from your theory which are subsequently confirmed by experimentation. So far, you have found no evidence which falsifies your theory. Then someone else comes along and says you're wrong. He says that it is obvious a demon causes disease B. You ask how he would test his theory and he tells you that he would test his theory by requiring that you show to his satisfaction that germ B causes disease B. If you can't show this to his satisfaction, he will conclude that your theory has been falsified and that his theory has been confirmed.

It cannot be emphasized enough that intelligent design theory is scientifically useless. As mentioned before, it doesn't explain anything. The intelligent design "explanation" of complex adaptation is that an even more complex designer designed the adaptation. But if you can simply postulate a complex designer, why not just postulate that eyes and bio-chemical pathways can magically appear on their own? More importantly, intelligent design doesn't lead to any predictions which can be empirically tested. Everything in nature is the way it is because the intelligent designer made it that way. Intelligent design theory hasn't led scientists to any new discoveries. No new facts have been ferreted out by the idea that living things are intelligently designed.

On the other hand, evolutionary theory has generated predictions which have led to countless new discoveries about nature. In virtually every issue of scientific journals which deal with evolutionary biology, there will be papers describing empirical tests of hypotheses generated by evolutionary theory. Sometimes the hypotheses are confirmed and some-times they aren't. The point is that the theory that organisms are the result of a long history of natural selection has enormous implications for what you might expect to find in nature.

One example which illustrates this point is infanticide. During the 1970's, anthropologist Sarah Hrdy studied Hanuman langurs in north-ern India. These monkeys have a mating system in which one male has exclusive sexual access to several females as long as he can drive other males away. Inevitably, however, a stronger male will come along and kick the harem owner out and take possession of the females in the group. The new male then begins to systematically hunt down and kill any unweaned (still nursing) infants. Females resist these attacks and often have help from close relatives, but the male usually succeeds because he is much bigger and stronger. Once the infants are dead the females stop producing milk and begin ovulating again. They commence mating with the infanticidal male.

Hrdy interpreted this systematic infanticide as an adaptation which increased the males' reproductive success. Males can control harems for only a limited time. Those males who killed infants sired by other males and thereby brought the females back into estrus sooner had greater reproductive success down through the generations than males who did not kill unweaned infants.

Hrdy's idea was initially met with emotional denunciations. This couldn't really be happening. These couldn't be normal monkeys. They must be crazy. After all, killing infants in this way is bad for the species. It couldn't possibly be a normal, adaptive behavior. How could you find out?

The idea led to a prediction. Similar infanticide should be found in other species where conditions would make it pay off with greater genetic success. In other words, in situations where males could selectively identify offspring that were not their own, where males were larger and stronger than females, and where females remained under the control of the new dominant male (thereby having no other mating options), then infanticide should have evolved. Once scientists started looking the results came pouring in. This kind of "stepfather" infanticide is widespread in mammals (gorillas and lions being two examples). In fact, infanticide is common throughout the animal kingdom. It is committed not only by unrelated males, but also by rival mothers and mothers themselves under certain circumstances.

This is just one of a multitude of new facts which have been discovered as a result of evolutionary theory. The point of this exercise is to contrast evolution by natural selection with intelligent design. What would an intelligent design theorist have made of infanticide in langurs? Would he have even reported it? How would he have interpreted it? For some unfathomable reason, the designer has designed langur males to kill infants that are not their own. What is the designer's "purpose?" (We'll return to this interesting question later.) Would an intelligent design biologist have come up with any predictions from his observations of langur infanticide?

I have been emphasizing that intelligent design theory doesn't explain anything, doesn't predict anything, and is completely untestable. The response of intelligent design advocates to a peculiar set of phenomena further drives home these points. Living things display a wide variety of exquisite adaptations which justifiably earn the admiration of both evolutionists and creationists. However, they also exhibit a multitude of stupid features which support the idea that they are the result of evolution by natural selection: a historical, trial and error process which can only work with what is currently available. In other words, it can only modify what has gone before. It can't redesign from scratch.

There are many examples. One classic is the fact that the retina in vertebrate eyes is inverted, that is, the photoreceptive cells (rods and cones) are at the back of the retina instead of in the front. This results in an absurd situation in which the light has to travel through blood vessels and nerves before it reaches the rods and cones. (Normally we are not conscious of these blood vessels in front of the receptor cells. However, you may have seen them during a trip to the ophthalmologist. When the room is completely dark and a light is shined in your eye at just the right angle you can see them. I remember being slightly startled at seeing a ghostly net of blood vessels the first time I experienced this.) It also results in a blind spot on the retina where all the nerve fibers gather into the optic nerve and exit the eye on their way to the brain. Another problem with this arrangement is that it leaves us vulnerable to detached retinas. We know that this strange setup is not the result of some unknown design necessity because cephalopods (squids and their relatives) have eyes almost identical in design to vertebrate eyes yet their retinas have the receptor cells in front. I don't know if there is any data on whether cephalopods have superior visual acuity to vertebrates or not. The point is simply that the cephalopod eye shows us that it is possible to have a functional retina without the seemingly absurd inversion seen in vertebrate eyes.

Another well known example is the crossing of respiratory and digestive tracts in vertebrates. Because of this crossing all vertebrates are capable of choking to death. Humans are particularly susceptible because of our descended larynx.

Human childbirth provides another example. I can't imagine that any woman would consider childbirth an intelligently designed process. Why force a baby out through the pelvis when it would be much easier to have an opening through the belly? The pain and danger of human childbirth is the result of a historical legacy and the tradeoff between conflicting selection pressures. The historical legacy is the fact that in our ancestors the birth canal passed through the pelvis. Selection for bigger brains made passage through the pelvis during birth a problem. One solution would be to widen the pelvis, but that hinders the efficiency of bipedal walking. If natural selection could start from scratch it might have produced an opening through the belly and avoided the tight squeeze through the pelvis. Since selection can't redesign from scratch the compromise result is that human infants are essentially squeezed out early in a more helpless state than other primates.

The bizarre appearance of flounders betrays their history. They are descended from fish that originally had the typical laterally compressed (like a vertical knife blade) shape of bony fish. Lying on their bellies on the bottom was not a viable option during their evolution. The path their ancestors followed involved lying on their side to flatten themselves against the bottom. This led to the strange asymmetry of their eyes as the eye that was aimed at the bottom slowly rotated around the head through the generations.

We could go on and on with other examples of "non-intelligent" design, but I will mention only one more. The recurrent laryngeal nerve (the fourth vagus nerve) is one of the cranial nerves which originally evolved in fishlike ancestors. In fish, this nerve passes behind the sixth gill arch. This is a direct path from the brain to the gills. During the evolution of mammals these gill arches have been substantially modified. The sixth gill arch became the ductus arteriosus, a connection between the pulmonary artery and the aorta which exists only in the fetus (lungs are non-functional in the fetus and this connection causes the blood to largely bypass them—oxygen is supplied through the placenta). The recurrent laryngeal nerve still passes behind this modified "gill arch." In other words, a nerve which goes from the brain to the larynx has a very circuitous route. It doesn't proceed directly from the brain to the larynx. Instead, it runs down the neck into the chest, loops around the aorta, and then goes back up the neck to the larynx. This detour is strange enough in humans, where it amounts to a little more than a foot of extra length. Now consider a giraffe's recurrent laryngeal nerve. You've guessed it. The nerve dutifully runs from the giraffe's brain all the way down its neck into the chest, loops around the aorta, and then back up the neck to the larynx. This stupefying detour requires the growth of several feet of additional nerve and no doubt slows transmission speed significantly.

So how do intelligent design theorists respond to these sorts of anomalies? Their responses are particularly illuminating concerning the testability issue. You see, Johnson, Behe and other intelligent design advocates want to have it both ways. Their human minds can immediately spot design, but when other human minds point out strange anomalies these are dismissed because the creator is "inscrutable," "whimsical," and "mysterious." He may have "multiple motives." "Clearly, designers who have the ability to make better designs do not necessarily do so." We cannot psychoanalyze the designer because "…the reasons that a designer would or would not do anything are virtually impossible to know unless the designer tells you specifically what those reasons are."

I think it is fairly obvious that intelligent design is not testable. Intelligent design theorists have no positive evidence and so they must rely solely on attempting to tear down evolution by natural selection and hoping that people fall for the bogus logic that this somehow proves their case. Yes, science hasn't explained everything. Evolutionary biology has a lot of work to do, but unexplained doesn't mean unexplainable. Scientists will keep looking for answers. Intelligent design theorists, on the other hand, won't even bother to look for answers. That's why intelligent design is not only useless but dangerous to science.

Human Evolution

However much Johnson and other creationists may mention arguments that do not directly involve *Homo sapiens*, it is clear that it is the evolution of humans that is most troubling to them. I'm sure it wouldn't bother them at all if everything else in the living world had evolved, but humans had been separately specially created by God for a purpose. This attitude was illustrated by a letter to the editor in the *Tulsa World* in which a parent complained about the terrible harm inflicted on his child by a sign at the zoo's chimpanzee exhibit which stated (correctly) that humans and chimps had a common ancestor. His objections apparently paid off since the exhibit no longer contains the offending reference.

This is a truly sad situation. I am always amazed at adults who insist that kids must be "protected" from evolution. The usual argument given is that if you teach kids about evolution they will immediately plunge into immorality because if evolution is true, and we're "just" animals, then everyone will be out raping, stealing, killing, and so on. Johnson clearly thinks that evolution and naturalism are the root of all evil. We'll return to this later. For now, what about the claim that humans and chimps share a common ancestor? Did humans and chimps really evolve?

Humans and chimps are extremely close biochemically. 98.4% of the "letters" in the DNA of humans and chimps is identical. DNA studies indicate that among the great apes chimpanzees and bonobos (sometimes called "pygmy chimpanzees" even though their size range overlaps that of "common" chimps) are most similar to humans at the genetic level, followed by gorillas and orangutans. Even more interesting is the fact that chimpanzees and bonobos *are more similar at a genetic level to humans than they are to gorillas.* In other words, humans are equally closely related to chimpanzees and bonobos, and more closely related to the two chimpanzees than chimps (or humans) are to gorillas. As Jared Diamond has noted, humans could be thought of as the "third chimpanzee," and would be classified as such by an extraterrestrial zoologist. Humans are African apes.

Another stunning fact that has come out of the molecular data is that all humans are incredibly similar at the genetic level. It turns out that there is far more genetic variation in chimp and gorilla populations living in a small geographic area than in the most distantly related humans. There are six billion humans spread out all over the globe and yet there is less genetic variation among humans than in a localized chimp population in Africa. This has interesting implications for human evolution and we'll get to this topic later.

Biochemical differences accumulate between non-interbreeding lineages over time because of independent mutations which take place in each lineage. Scientists can use "molecular clocks" to estimate when two lineages diverged from each other. These clocks are calibrated by taking a well documented divergence in the fossil record and using that date to estimate the necessary average rate of change to produce the present biochemical differences. Among the great apes orangutans split off the evolutionary line first, followed by gorillas and humans, with the split between chimpanzees and bonobos occurring last. Estimates vary, but the time of divergence for the chimp/human lines is between five and seven million years ago.

This is pretty straightforward stuff. So what does Johnson object to? He accepts the close biochemical similarity between chimps and humans. He does not dispute that mutations take place. Well, if mutations take place, then biochemical differences must accumulate over time if two lineages do not interbreed. If you go back in time, you will reach a point where the lineages converge. What Johnson says is that all the connecting intermediates are merely hypothetical and that scientists are *assuming* that they existed because scientists have a prior commitment to common descent.

But scientists are not assuming common ancestry. What they are "assuming" is the common sense fact of life that all animals have parents, grandparents, and so on. No one has ever observed a human or chimp or any other organism spring into existence from nothing. Once you accept this fact (and anyone who doesn't is just being silly) and the facts discussed above, any rational person will *conclude* (not assume) that chimps and humans had a common ancestor.

Incidentally, Johnson uses this same tactic to dismiss all the molecular evidence which cries out that evolution occurred. Here is his preemptive strike: "It will not be enough to find that organisms share a common biochemical basis, or that their molecules as well as their visible features can be classified in a pattern of groups within groups." He says this, of course, because the data *do* show that organisms share a common biochemical basis and that classification by molecular similarity matches extremely well with classification by visible characteristics. The fact that molecular family trees closely match family trees constructed from visible features is powerful evidence for evolution. As with the case of fossils, it is important to remember that things could have turned out very differently. It could have been that molecular family trees bore no resemblance at all to morphological family trees. It could have turned out that living things did not share a common genetic code (the three-letter "triplets" in DNA which code for various amino acids in the construction of proteins). It could have turned out that human DNA was more similar to horse DNA than chimpanzee DNA. For that matter, it could have turned out that humans had a totally different molecule of heredity from all other living things. It might have been that animals were radically different from plants at the molecular level. None of these things happened. The basic oneness of life at the molecular level is amazing. Looking at life at the molecular level has been a boon to biologists in constructing patterns of descent. The fossil record is incomplete, and you can never know whether fossils had descendants, but you can be sure that molecules like DNA and proteins had ancestors.

Johnson constantly implies that the molecular data are not evidence for evolution, because they do not demonstrate the existence of ancestral intermediates. You have two choices here. You can assume the common sense fact of life that organisms don't appear out of nothing and conclude that they shared a common ancestor. This choice explains the data perfectly and does not require that you invoke any unknown or unseen mechanisms. Or, you can claim that there is no common ancestry. This choice does not explain the data and requires that you invoke an unknown, unseen mechanism. Furthermore, this mechanism has to have operated millions of times in the past as each species was independently popped into existence, even though the mechanism has never been observed in the real world. In the particular case of chimps and humans, the mechanism also put a similar pseudogene (nonfunctional gene) in exactly the same location in the genomes of each species.

This last point brings out another compelling argument for common ancestry. Fittingly for our purposes, it involves the legal profession. John Avise discusses the argument in his book, *The Genetic Gods*. When lawsuits are brought in plagiarism cases, the question is always whether similarities are the result of "separate creations" or copying from "ancestral texts." One thing which courts look for in deciding this question is what Avise calls "mistakes of detail." When mistakes of detail appear in both texts, the court concludes that plagiarism has occurred, since it is extremely unlikely that both authors would have made the identical detailed mistake(s).

It turns out that the genomes of living things are full of "mistakes." These errors are often shared by similar species. Pseudogenes are one type of such errors. Pseudogenes are genes which are similar to known functional genes but which have been rendered nonfunctional by mutations. They are "gene corpses." Avise explains:

These gene corpses often are shared by and display fine details of similarity across species suspected to be related. For example, both humans and great apes possess a pseudogene related to a functional gene encoding immunoglobulin epsilon (an antibody protein involved in allergic responses). Furthermore, the pseudogene appears in precisely the same location in the genomes of humans and chimpanzees, an exceedingly improbable outcome under a hypothesis of independent origins. Another example involves pseudogenes in the nucleus apparently transferred from the mitochondrion. Today, these "fossilized," nonoperational nuclear copies of mitochondrial genes are carried by several primates, including humans.

These shared mistakes in genomes are easily explained by common ancestry. The pseudogenes arose at some point in the history of primates and were simply copied down the generations to descendants. If intelligent design theory were true, it would mean that an incompetent designer made the same detailed errors over and over as he independently flashed each species into existence. Any judge who did not have a supernatural bias would conclude that the species in question were the result of descent with modification. (Incidentally, it is evidence such as this which leads Michael Behe to accept the common ancestry of apes and humans. The argument here is only with Johnson's version of intelligent design.)

Let's get back to Johnson's claim that the intermediates connecting humans and chimps to their common ancestor are merely hypothetical. Consider your female (using one sex for convenience) ancestors. For argument's sake, let's use a generation time of 20 years. If we go back 1,000 years, that would be 50 ancestors. Can you identify all of them? I suspect not, but would you say they were hypothetical? Go back 10,000 years. Now we have 500 ancestors. Would you deny that any of them existed? Would you deny that there was an unbroken chain of parent-child links? At what point do your ancestors become hypothetical?

Now put on your common sense thinking cap. You know that no one has ever seen or demonstrated that animals like humans or chimps appear out of nowhere without parents. You know that the only way we can do any kind of historical science is to assume that the same laws of nature applied in the past as they do today. What happens as we trace our ancestors back further in time? Here it is useful to refer to a revealing mental exercise proposed by Richard Dawkins in his article entitled "Gaps in the Mind." Dawkins is here attacking what he calls the "discontinuous mind," an affliction from which Johnson suffers greatly.

Dawkins has us recall events in which large numbers of people hold hands to form long human chains (across America, for example) in order to benefit some cause or charity. Imagine, he says, arranging one across the human home continent of Africa, beginning on the Somali coast and proceeding westward along the equator into Kenya. You stand on the ocean's shore in Somalia facing north, with your mother's right hand held in your left hand. She in turn holds the hand of your grandmother, who holds the hand of your great-grandmother, and so on. Allow one yard per person as the chain goes off to the west. How long will it be before we reach the common ancestor of humans and chimpanzees? For argument's sake, assume 20 years per generation and six million years back to the common ancestor. The answer is about 170 miles. We are not even halfway across Kenya.

Now the common ancestress turns and faces east. In her right hand she holds the hand of the daughter from whom we are all descended. With her left hand she then grasps the right hand of her other daughter (sticking to females for convenience), from whom all chimpanzees are descended. The two sisters are facing each other. A second chain is then formed as the second daughter holds her daughter's hand, who holds her daughter's hand, and so on, all the way back to the coast. Sister faces sister, first cousin faces first cousin, and so on back to the beach where you face your chimpanzee cousin.

Dawkins now has us imagine walking up the line like an inspecting general. All along the way mothers will resemble their daughters as much as they do today. They will love their daughters just as they do today. There are no sharp discontinuities anywhere along the line. There may be "punctuations" in the sense of greater overall change during some miles than others, but nothing you would notice at the generation-by-generation level. By the time you have walked about six miles, you would have passed the common ancestor of all modern humans. After about 15 miles you might think these people look a little different but you can't quite put your finger on it. At 40 miles, you've decided that these people are definitely somewhat strange. Paleontologists might call them *Homo erectus* (or *Homo ergaster*), but you couldn't point to any obvious break in the chain where *sapiens* became *erectus*. As you continued your extensive hike, you would eventually walk past other people (creatures?) which paleontologists might call *Homo habilis* and then some they might call *Australopithecus*. But there would be no point at which you could draw any defining lines. All you could say is that the people/creatures/apes at 150 miles inland are definitely different from the people on the beach.

As you reach the common ancestress, you note that there is nothing special about her. Her daughters look just as much like her as all the other daughters in the chain look like their mothers. Now you turn and begin inspecting the chimpanzee line as you walk back to the coast. Again you see no distinctive gaps along the way. However, you have no labels to attach to the people/creatures/apes on this side because (assuming no misinterpretation) no fossils have been found. (This is no doubt a source of great comfort for chimpanzee creationists.)

Johnson, of course, would say that there is no way that we can know that these ancestors existed. We are simply assuming, because of our inherent naturalistic bias, that these hypothetical beings existed. Here is a more interesting way to think about it. Given that we know that the "letters" in the DNA of chimps and humans are over 98% identical, that mutations occur, that animals do not come into existence without parents, and that biochemical differences must therefore accumulate between non-inter-breeding lineages over time, *there would have to be supernatural interven-tion for humans and chimps not to have a common ancestor.* Johnson, of course, *is* claiming that the supernatural is involved somehow, but he has-n't supplied us with even a scrap of evidence.

What about the fossil evidence for human evolution? The hominid—hominid refers to any animal on the human side of the split from the last common ancestor with chimps—fossil record is reasonably good. One way you can tell that there are a fair number of fossils is that paleoan-thropologists have plenty to argue about. Another way that you can tell that the record is pretty good is that Johnson spends a fair amount of time in *Darwin on Trial* and *Defeating Darwinism* complaining about what a terribly subjective field paleoanthropology is. He feels that it is highly probable that physical anthropologists are only seeing what they want to see. He is therefore skeptical that any of the "alleged hominid species" pro-vides evidence of human evolution. Still, he tells us that he is willing to assume for the sake of argument that upright-walking australopithecines existed, and that there "may also have been an intermediate species (*Homo erectus*) that walked upright and had a brain size intermediate between that of modern men and apes." If we make these assumptions, he says, there are possible intermediates between apes and humans, but no smooth line of development. We would still have no mechanism for how these transformations could have occurred. In other words, he doesn't think the fossil interpretations are reliable, but even if they are reliable, he will invoke the Argument from Personal Incredulity to show that the changes could not have been accomplished by evolution.

Fortunately, you don't have to assume anything about the hominid fos-sil record. The oldest hominid fossil for which any information is available is called *Ardipithecus ramidus.* It has been reported as 4.4 million years old and very chimplike, although a complete description has not been pub-lished yet. This species is probably very near the chimp/human split. As scientists look in older sediments they will probably find fossils which could be chimp ancestors or human ancestors, or both. They won't be able to tell for sure, but that's the point. Any fossils they find close to the branch point will be part of a confusing continuum. No one knows what the chimp/human ancestor looked like, but fossil evidence indicates that it was probably more similar to modern chimps than modern humans.

Hominid fossils slightly over four million years old have been assigned the name *Australopithecus anamensis*. This species is less apelike than *Ardipithecus ramidus* and the anatomy of its tibia (the big bone in the shin) indicates that it walked upright.

Fossils assigned to the species *Australopithecus afarensis* have been found ranging in age from about four to three million years ago. The famous fossil named "Lucy" was a member of this species. This species was an upright walker with a very apelike head. Its limb proportions were also apelike. You may wonder how scientists can infer that a fossil animal walked upright. There are many clues. The foramen magnum (the opening in the base of the skull through which the spinal cord passes) is located near the center of the base of the skull in upright walkers (as opposed to nearer the back of the skull in quadrupeds). Vertebrae are different in bipeds. The spine has an S-shaped curve to it. The pelvis is shorter and more bowl-shaped in upright walkers. This allows more efficient muscle attachments as well as helping to hold the internal organs. The femur (thigh bone) has a longer "neck" attaching it to the pelvis. This allows it to angle inward with a slight bend at its lower end which squares it up with the top of the tibia. This brings the center of gravity of the body more directly over each foot during walking, substantially improving efficiency. Apes have a distinctive waddle when they try to walk upright (imitated by the actors in *Planet of the Apes*) because their femurs do not angle in. The waddle results because with each step their center of gravity is not above the foot on the ground. They tip inward more than humans do until the other foot catches them and gravity tilts them back the other way.

In Tanzania, there is an extraordinary site called Laetoli. A truly remarkable discovery was made there in 1978. Preserved in a layer of volcanic ash 3.6 million years old was a series of hominid footprints about 30 yards long. Two hominids (and many other animals) had walked through the area just after a volcanic eruption of ash followed by a light rain which rendered the ash mushy. The ash subsequently dried and hardened in the sun. Another layer of ash from the volcano preserved the footprints until they were uncovered 3.6 million years later. It is not known for sure which species of hominid made the tracks, but *A. afarensis* is the likely suspect since its bones are the only hominid fossils found in the area. Paleontologists may argue over whether australopithecine bipedalism was exactly like modern human walking, but it is ridiculous to imply (as Johnson does) that there is some question over whether they were bipedal at all. They were definitely bipedal by 3.6 million years ago, and good evidence that they were bipedal at least four million years ago.

The first hominid fossil found west of the Great Rift Valley in Africa was described in 1996. It was discovered in Chad and was named *Australopithecus bahrelghazali*. It is estimated to be 3-3.5 million years old.

At about 2.5 million years ago, things get really interesting. Anthropologists have named several different species from about this time: *Australopithecus africanus, aethiopicus,* and *garhi. A. garhi* appears to have butchered meat and may be associated with the first extensive use of stone tools. *A. aethiopicus* was a "robust" australopithecine. These hominids are noted for their massive jaws, teeth, and chewing muscles (as evidenced by the crests on the top of their skulls to which chewing muscles were attached). They apparently ate a lot of tough plant material. Subsequent robust species included *A. robustus, boisei,* and *crassidens.* The robust australopithecines did quite well but appear to have gone extinct about one million years ago.

The first examples of fossils that anthropologists put in the genus *Homo* are about 2.3 million years old. *Homo habilis* and *Homo rudolfensis* are just under two million years old. *Homo ergaster* is slightly more recent. There is controversy over whether this species is the same as *Homo erectus*. A spectacular skeleton of *ergaster/erectus* dated at 1.6 million years old was discovered in 1984. Called the "Nariokotome boy," it is a nearly complete skeleton from an early adolescent. The boy had already grown to over five feet in height and probably would have reached six feet had he lived to adulthood. His skeleton indicates he was an efficient upright walker. Brain size in *ergaster/erectus* is intermediate between australopithecines (and chimpanzees) and modern humans. *Homo erectus* is the first hominid known to have migrated out of Africa. *Homo erectus* groups may have left Africa as early as 1.8 million years ago. They eventually reached as far as Europe, China (Peking Man), and Java (Java Man). They may have survived in Java until as recently as 27,000 years ago.

Homo erectus appears to have given rise to *Homo heidelbergensis* about 600,000 years ago. This species was slightly more modern looking than *erectus*. *Homo heidelbergensis* is thought to have given rise to *Homo neanderthalensis* (Neandertal Man) in the Middle East and Europe, and *Homo sapiens* (us) in Africa between about 200,000 and 150,000 years ago. Neandertals were very powerfully built with big brow ridges and somewhat flattened craniums containing brains slightly larger than ours. *Homo sapiens* migrated out of Africa about 100,000 years ago and subsequently spread around the world, apparently displacing the Neandertals and any relict *Homo erectus* populations that still existed. The displacement may have been violent (given the human record in dealing with previously unknown groups or cultures) or the other species may have slowly faded away as the result of ecological competition. Neandertals and modern humans coexisted for tens of thousands of years before the Neandertals went extinct a little less than 30,000 years ago. We are currently the only living hominid species, but at various times in the past there were several different hominid species living at the same time. (I have a soft spot for the various legendary "wildmen" said to live in various parts of the world. I would be delighted if any of them turned out to actually exist. Alas, so far there is no convincing evidence that they do.)

I should add that the scenario I have just given is the "Out of Africa" scenario for modern human origins. No one disputes the migration of *Homo erectus* out of Africa over one million years ago. However, the "multiregional" theory holds that modern humans did not migrate out of Africa relatively recently and displace the other hominids. Instead, it maintains that *Homo erectus* populations that had already migrated out of Africa to various parts of the world slowly evolved separately into *Homo sapiens* with enough gene flow taking place between the different regions to keep everyone a single species. Given the genetic data I mentioned near the beginning of the chapter, the "Out of Africa" scenario makes a lot more sense to me. Humans are so monotonously similar at the genetic level that scientists believe that our ancestors must have gone through one or more population "bottlenecks" during which much genetic variation was lost as the population was reduced to a few thousand people. The genetic data also point to the fact that the small population we are all descended from lived in Africa. (This is inferred by constructing a genetic tree using amounts of genetic difference in the same way as the ape tree mentioned earlier. In the case of modern humans some Africans branch off first, followed by other Africans, followed by some more Africans, followed by everybody else in the world. The root of the tree is in Africa).

I have gone through this grossly oversimplified whirlwind tour of the human fossil record to give you some idea of the data that are available to paleoanthropologists. I compiled it from several sources (see the notes) and I encourage you to pursue the interesting details yourself. For the real skeptics, I particularly recommend Donald Johanson and Blake Edgar's book, *From Lucy to Language*. This book is worth consulting just for the pictures. There are beautiful actual size photos of many of the most important fossils, including almost all of the ones mentioned above. Also included are photos of chimpanzee and gorilla skulls. You can see for yourself whether the anthropologists are as biased as Johnson implies they are.

Johnson, of course, talks about gaps in the record and that there is no proof that any of the purported hominid fossils is actually an ancestor of modern humans. The issue of gaps is tiresome by now, but it is strictly true that you can never tell whether a specific fossil had descendants. But, as we have seen, there are many fossils between 4.4 million years old and the present which exhibit characteristics intermediate between humans and other apes. Each individual fossil may or may not have had descendants, but some of their contemporaries surely did or we wouldn't be here. Creationists try to lump all fossils into the category of "ape" or "human" and imply a chasm in between, but we've seen how ridiculous this tactic is. Paleontologists are forced by Linnaean classification to give a fossil a name. This in itself creates "gaps." If it's called *Australopithecus*, it must be an ape, right? And if it's *Homo*, it must be human. Always think in terms of individual animals and parent-child connections and you won't be deceived by this tactic. Think back to the ancestral chain in Dawkins's mental exercise. There is no point in an ancestral chain where you can draw a non-arbitrary line and say that you have gone from one species to another. People get confused on this point. They can't see how one species can change into another. Picture the history of life as a great, branching tree. As long as you stick to currently living organisms, Linnaean classification is not too confusing. Likewise, if you took horizontal time slices through the tree, the cutoff twigs could be classified fairly easily into discrete species. As long as you stick to time slices the confusing intermediates connecting species to each other are all in the past (except for "ring" species—see the notes). They run back in time from each species until they converge at the branching point. If you don't stick to time slices, if you move continuously along a branch through time, then dividing a lineage into different species becomes arbitrary and confusing. The problem for paleontologists is that they don't deal with time slices. They find fossils from all over the tree. Scientists find fossils of different ages and assign them to species based on their physical characteristics. When dealing with the past, Linnaean classification inevitably causes confusion by imposing discrete categories on biology's continua. Even with the few fossils they have, scientists sometimes have tremendous arguments over which species a particular fossil should be assigned to. Imagine the arguments they would have if they had fossils from every link in our ancestral chain.

Consider another calculation concerning gaps in the record (similar to the one done by Richard Dawkins that was mentioned in a previous chapter). This time we'll just go back to the chimp/human common ancestor. Using the previous assumptions (daughter lies down and dies on her mother's grave and is subsequently fossilized, each fossil is compressed to one centimeter in thickness, 6 million years back to the common ancestor, 20 years per generation), how much rock would we need to have a continuous, gapless fossil record clear back to the chimp/human split? The answer is about 1.9 miles, which is about double the amount of vertical section exposed in the Grand Canyon, and that section spans hundreds of millions of years. But to do the kind of unbiased analysis that Johnson would no doubt require, we will probably need uncrushed skulls so let's allow 6 inches for each fossil. The answer is now about 28 miles. Paleoanthropologists must be drooling.

It is also important to realize that although a real generation-by-generation chain of ancestors connects us to our common ancestor with chimpanzees, the people in the chain of ancestry were, of course, not the only people alive at the time, nor was our line the only line of hominid evolution. As we have seen, we are the only hominid species alive today, but at various times in the past there were at least three coexisting. Indeed, Johnson comments on the fact that modern humans, Neandertals, and *Homo erectus* probably coexisted in various parts of the world up until about 30,000 years ago. (In the "Out of Africa" theory discussed above modern humans migrated out of Africa about 100,000 years ago. They definitely encountered Neandertals and may have encountered *Homo erectus* in Asia.) He suggests that, "If they could interbreed, then it would be more accurate to say that they were all a single species, *Homo sapiens.*" You're supposed to get the hint that if they were all one species then there must not be any evolution involved (or that the three different forms are just the result of microevolution within *Homo sapiens*). But species are fuzzy things. Many species can interbreed (sometimes the hybrids are sterile and sometimes they're not). Horses and donkeys can. Lions and tigers can. Chimpanzees and bonobos can. It's at least possible that humans and chimpanzees could interbreed. Somehow I doubt whether Johnson would decide that humans and chimps were the same species if such an event were to occur.

At any rate, the record of hominid evolution is not a single lineage leading to us. Many other bipedal hominids existed which did not make it to the present. Their lines would have joined the ancestral chain in our mental exercise at various places between the coast and the chimpanzee/human ancestor. They just didn't make it to the beach.

Another Johnsonian (and standard creationist) tactic is to harp on disagreements between paleoanthropologists and to insinuate that this means that the fact that humans evolved is somehow in question. Paleoanthropologists do disagree about many things. Are these fossils in the same species or not? Are the differences due to sexual dimorphism in one species or are they due to differences between species? Is this species a descendant of that one? There are many different fossils and different possible evolutionary relationships. Scientists will continue to argue about such things and views will continue to change as new fossils are found, but this doesn't mean that humans didn't evolve.

The fossil record indisputably shows that there once existed, between about 4.4 million years ago and the present, a slew of bipedal hominids. The ones nearer the 4.4 million year mark are very apelike. As you move toward the present, some of the fossils get more and more humanlike. Ultimately, there are fossils which are indistinguishable from modern people. If humans didn't evolve, how do you explain the fossils and the molecular data we've covered in this chapter? As we've seen, Johnson avoids like the plague giving any specific alternative to evolution. Once again, you need to ask yourself why. The intelligent designer must have been pretty busy creating all those hominids separately and timing their miraculous appearances so that it looks just like an evolutionary sequence.

Finally, remember that Johnson is fond of saying that the fossil record doesn't fit with Darwinian predictions. Consider this prediction, made in the 1870's by Darwin himself. He predicted that the ancestors of modern humans, the intermediate forms between humans and apes, would be found in Africa. He based this on the similarities between humans and African apes. Think carefully about that prediction. Darwin was essentially saying that if you go look in the rocks of Africa you should find fossils which have a strange mix of human and ape characteristics. This, of course, is exactly what happened. Contrast this with Johnson's "theory." It doesn't predict anything. As he says, "There is a whole lot of evidence out there, and even a false theory is likely to be supported by some of it." How true. The problem is that he not only hasn't provided us with *some* evidence for his theory, he hasn't provided us with *any* positive evidence.

Johnson's Utility Function

We've waded through naturalism, micro- and macroevolution, fossils, intelligent design, and human evolution. Now I want to go into an area that Johnson mostly evades or ignores. What is this thing that he wants to avoid talking about? It's the real world. If God (or whoever the intelligent designer is) created the living world for a "purpose," if things really are designed, then we should be able to figure out from the creation what that purpose or design was. We have already seen how Johnson, Behe, and other intelligent design advocates want to have it both ways. They maintain that design is blatantly obvious, but then say that there is no way we can know what the designer's motives were. This won't work. If human minds can see such "obvious" design, then human minds should be able to see what the design is for. Johnson would readily say that wings are designed to fly and eyes are designed to see. Why, when asked about the whole creation, does he say only that it is for a "purpose?"

Richard Dawkins, in *River Out of Eden*, considers this question in detail. In a chapter entitled "God's Utility Function" he imagines a Divine Engineer who designed the living world. (Johnson quotes Dawkins from this chapter, but he never addresses the main question.) What was the Divine Engineer trying to maximize? "What was God's Utility Function?" Dawkins then gives several possible answers (my favorite being: "to maximize David Attenborough's television ratings") before arriving at the true one: maximizing DNA survival. George C. Williams, in *The Pony Fish's Glow*, goes through a similar exercise and reaches the same conclusion:

> ...a system in which the ultimate purpose in life is to be better
> than your neighbor at getting genes into future generations, in
> which those successful genes provide the message that instructs
> the development of the next generation, in which that message
> is always "exploit your environment, including your friends and
> relatives, so as to maximize our (genes') success," in which the
> closest thing to a golden rule is "don't cheat, unless it is likely to
> provide a net benefit"...

In other words, organisms are "designed" to maximize the transmission of their genes into the next generation, both through their own survival and reproduction and by aiding the survival and reproduction of relatives (who contain copies—in various proportions, depending on the degree of relatedness—of the same genes). At times, genes get themselves transmitted by promoting unselfish behavior at the level of the organism, but this is always a result of selfishness at the genetic level. Much of the time, behavior is overtly antagonistic. There is no evidence for any force acting for the good of the species or ecosystem.

Dawkins and Williams go through multiple examples to illustrate their point including sex ratios, sexual competition, predation, parasitism, aging, cannibalism, and infanticide. Reviewers of Johnson's books have brought up a similar point. David Hull noted that "the evolutionary process is rife with happenstance, contingency, incredible waste, death, pain, and horror." And later in the same review, "The God of the Galapagos is careless, wasteful, indifferent, almost diabolical."

How does Johnson respond? He calls this a "Darwinist doctrine of God." Wrong. The world is as it is. The world is full of suffering and all the horrors mentioned above whether or not evolution or creation is true. Darwinian theory does not imply a diabolical Creator. It *explains* why the world is the way it is. Intelligent design theory explains nothing and *does imply* a diabolical Creator.

Let me be clear. Evolutionary biology, like all sciences, has nothing to say about God. It is not necessary to involve God to explain adaptations or the history of life just as it is not necessary to involve God to explain chemical reactions or relativity. Remember, it is Johnson and other intelligent design advocates who are pulling God into science by proposing intelligent design as a *scientific* theory. His theory, vague as it is, seems to be that God separately created all living things for a "purpose." As we've already seen, he believes that God "left his fingerprints all over the evidence." If that's true, we should be able to look at the natural world and discern what that purpose is. Johnson maintains that there is no entirely satisfactory explanation for the "problem of suffering" (he also seems to be concerned only with human suffering). But suffering is only a problem (I speak here of a metaphysical problem—obviously everyone is distressed at the suffering in the world) for proponents of intelligent design who are proposing as a scientific theory that God designed and separately created every extinct as well as every living species. (Actually, it's only a problem for Johnson's particular version of intelligent design, wherein the designer is a benevolent, omnipotent being. I know of no intelligent design theorists for whom the designer is not this traditional picture of God. For all I know, there may be intelligent design advocates who believe the designer made nature the way it is because he enjoys cruelty and suffering.) It is an indisputable fact that the vast majority of organisms die before reproducing because of spontaneous abortions, starvation, predation, disease, cannibalism, parasitism, infanticide, siblicide, and other gruesome causes. It seems to me that an honest intelligent design theorist should tell us just what the designer is up to. Again, it does no good to say that the designer is "mysterious" or "inscrutable." If you can see design, then you should be able to see what the design is for.

I emphasize again that evolutionary theory explains these phenomena. It does not imply that they are good, or that the world ought to be the way it is. From a human standpoint, these facts of life are distressing, but nature itself is neither good nor evil. It simply is.

It is easy for modern humans (at least in the industrialized, Western world) to overlook the suffering which goes on in nature. It is relatively difficult to die young in most parts of the United States. Before modern medicine and other advantages were available, however, the majority of humans died before reproducing just as is the case for all other animals in the wild today.

Consider the life of a gazelle (or whatever your favorite prey animal is). You are immediately a target for predators the day you are born. If you make it to adulthood (a very big if), you are still at constant risk from predators. You must compete (sometimes violently) for mates. If you get sick, or are weakened by parasites, or suffer an injury you can't go to the doctor or have the luxury of danger free recuperation time. You will be run down and killed if your running speed is impaired by any of these causes. Even if you do survive to adulthood your physical abilities will inevitably decline with age, but you won't get the luxury of a golden retirement. The gazelle equivalent of a 40-year-old human is running a race with predators that it will inevitably lose.

The plight of the predator is not much better. The same risks of disease, parasites, and injury remain. The same potentially bloody sexual competition remains. With age, it becomes increasingly more difficult to find and capture prey. Don't be misled by misguided nature programs which talk about the "balance of nature" or praise nature's plan for weeding out the weak, old, or sick. Picture your child or grandmother being slaughtered. Picture your sick aunt being killed for the good of the ecosystem.

I am belaboring the point, but I want to emphasize again that design advocates cannot be allowed to have it both ways. You can't say that design is obvious and then say that you can't say what the design is for. Evolutionary biology has put its cards on the table. Organisms are "designed" by natural selection to maximize the number of copies of their genes that get into the next generation. This idea has been enormously successful. It has almost limitless implications for what you might expect to find in nature. It suggests all kinds of testable hypotheses. It has found answers to some of the most interesting questions about life. If intelligent design advocates have evidence that this view is wrong, they should let us see it. If they think organisms are "designed" for something else, they should tell us. Johnson tells us that living things are as obviously designed as rockets and computers. We know what rockets and computers are designed for. If the design of living things is as obvious, he should tell us what they are designed for.

Consider again the phenomena (sex ratios, sexual competition, predation...) mentioned above. I said that evolutionary theory can explain them. I refer you to the books mentioned above and the notes for references. Does Johnson attempt to explain any of them? I could find only one instance in which he offered an alternative explanation to evolutionary theory for *any* phenomenon in nature. What was this pearl of wisdom? He attributes the elaborate, encumbering plumage of the peacock to a "whimsical Creator." Why? Because "...it seems to me that the peacock and peahen are just the kind of creatures a whimsical Creator might favor, but that an 'uncaring mechanical process' like natural selection would never permit to develop." He further informs us that "sexual selection is a relatively minor component in Darwinist theory today" when it is actually one of the most interesting and exciting areas of evolutionary biology. He also tells us the following:

> ...what I find intriguing is that Darwinists are not troubled by the unfitness of the peahen's sexual taste. Why would natural selection, which supposedly formed all birds from lowly predecessors, produce a species whose females lust for males with life-threatening decorations?

Here Johnson betrays his ignorance. The peahen's sexual taste *is* "fit" (adaptive) at the level of the genes. The gaudy plumage of peacocks is the result of a long history of peahens selecting the most elaborately decorated males. Any peahen which bucks the trend by mating with a drabber male will produce sons with a lower probability of attracting mates themselves and daughters which have the tendency to make the same mistake. It would be better for the species if females selected for more "utilitarian" features but it is better for the genes which produce the gaudy features if females select the gaudiest males.

How does the trend get started? The original preference for longer or more elaborate tails may be for utilitarian reasons (greater stability in flight, for example). It becomes an encumbrance only after "runaway" sexual selection. However, it does not matter whether the original preference makes "sense" or not. Even if it is totally arbitrary, once a majority of females have genes for that preference, those genes can best replicate themselves by building females who choose gaudy males. These females in turn produce gaudy sons and daughters who prefer gaudy males and so on.

Other sexual selection theories maintain that gaudy or bright features really do indicate underlying genetic quality (such as resistance to parasites or ability to survive *despite* having ornamental handicaps). It turns out that in the case of peacocks, peahens may not be making a purely arbitrary choice when they select the fanciest tails. Recent research found that offspring of males with "more elaborate trains" grew more rapidly and had improved survival rates. There is evidence to support both "runaway" and "good genes" theories of sexual selection.

But I'm getting far away from my main point. In the one case where Johnson tries to provide an alternative to evolutionary theory he gets it completely wrong. Natural selection is primarily about maximizing *genetic* success. It does not "care" about species, ecosystems, or the "balance of nature." It is primarily concerned with the differential survival and replication of genes. If that turns out to be bad for the species, that's just the way it is. There are many phenomena in nature which are bad for the species or individual that are readily explained and understood from the viewpoint of genic selection.

I mentioned infanticide in a previous chapter. Cannibalism is also common in nature. Siblicide (killing siblings) occurs in many species. Males compete—often violently—for mating opportunities, both before mating and after (sperm competition). Females exercise choice in mates and often manipulate fertilization after mating ("cryptic" female choice). The bizarre implements and tactics involved in insect mating alone would boggle your mind. None of these things makes sense in terms of "good of the species." They do make sense as the result of the struggle of individuals for reproductive success.

Amazingly, Johnson gives a brief summary of "selfish gene" theory in *Defeating Darwinism* and discusses both Dawkins and Williams (whose classic 1966 book *Adaptation and Natural Selection* cleared up a lot of sloppy thinking in biology and emphasized genic selection) but he clearly doesn't understand its implications. He also doesn't seem to have bothered to look at the natural world (or the scientific literature) to see if "selfish gene" theory is true. There is a huge body of evidence supporting it. It is difficult to concisely convey the range of mind-boggling things that have resulted from the competition of genes attempting to replicate themselves. (Here I must include the obligatory caveat that genes, of course, are not conscious and do not "try" to do anything. Genes are self-replicating molecules. Genes that are better at getting copies of themselves into the next generation inherit the earth at the expense of genes which are less proficient at doing so. It is convenient and has proven very fruitful for biologists to think and write about genes *as if* they are scheming to get themselves copied down the generations.) The "gene's eye view" of evolution has been spectacularly successful in explaining and discovering new facts about animal behavior. In addition, intragenomic conflict (conflict between genes in the genome of an individual organism) turns out to be widespread. I will mention a few phenomena briefly and encourage you to research the topics yourself.

Genes on the sex chromosomes (X and Y) can be at "war" with each other. The X chromosome often contains genes which are good for females but bad for males. The Y chromosome often contains genes which are good for males but bad for females. Genes on the X chromosome can actually spread themselves by damaging the Y chromosome. In many species genes have appeared on the X chromosome which cause the death of Y-carrying sperm. This results in males whose offspring are all daughters. The sex ratio of the species can become massively female skewed and put the survival of the species in jeopardy. This is obviously bad for the species, but good for the spread of the X genes in question.

Genetic imprinting is a phenomenon in which genes are expressed differently depending on which sex they passed down from. In mice (and probably other placental mammals) the placenta is made by genes from the father. The placenta aggressively invades the mother's blood supply and produces hormones which raise the mother's blood sugar and blood pressure. The mother responds with her own chemical countermeasures. In a "normal" pregnancy, a stalemate ensues and the baby is born with no complications. However, if the mother "loses" the battle with the fetus, she may suffer from high blood pressure or diabetes. If the fetus "loses" it may be spontaneously aborted. None of this seemingly senseless conflict would occur if there were no genetic conflicts of interest.

Johnson and other creationists like to picture the genome as an intelligently written affair. It certainly doesn't appear that way. Current estimates are that only about 3% of our DNA actually codes for proteins. The other 97% consists of so-called "junk DNA" or "selfish DNA." (Actually all DNA is "selfish" in the sense of the selfish gene mentioned above. It's just that the 97% of interest here doesn't code for anything useful to the organism. It's even more selfish.) Matt Ridley's book *Genome* (which delves deeply into intragenomic conflict) engagingly goes into the details of what all this junk is. Some of the junk has been inserted into the genome at various times by retroviruses. Some consists of the pseudogenes mentioned in another chapter. There are sequences about 20 "letters" long called "hypervariable minisatellites" (this is the DNA useful in forensic identification) which repeat themselves over and over at over a thousand different places in the genome. Strewn throughout the genome are various "genetic parasites" which do nothing for the organism. These parasites, as Ridley puts it, "exist for the pure and simple reason that they are good at getting themselves duplicated." Some of these parasites jump around in the genome, creating havoc when they land in working genes. The view that genes are selfishly "trying" to replicate themselves has been abundantly confirmed. How all this bizarre stuff fits into intelligent design theory is anybody's guess. Intelligent design theorists won't tell us.

Johnson wants you to believe that:

> Evolutionary biology is a field whose cultural importance far outstrips its modest intellectual and scientific content. Its sacred trust is to preserve the central, indispensable part of the modernist creation story, which is the explanation of how such things as life, complex organ systems and human minds could exist without a Creator to design and make them. We might say that the point of Darwinism is to refute the otherwise compelling teaching of Romans 1:20, which is that God's eternal power and deity have always been evident from the things that were created.

Silly me. I thought the point of Darwinism was to explain various phenomena in the biological world and to suggest further lines of research. I must have missed the lectures in college that went over refuting Romans 1:20 and maintaining the naturalistic conspiracy. Furthermore, if there were ever a field "whose cultural importance far outstrips its modest intellectual and scientific content," it is certainly intelligent design theory. We are still waiting for the first new scientific discovery about nature generated by the theory that living things are intelligently designed.

Incidentally, the pioneers of "selfish gene" theory have collected several Crafoord Prizes (given by the Royal Swedish Academy of Sciences to honor scientists in fields not covered by the Nobel Prizes) recently. W.D. Hamilton was honored in 1993, John Maynard Smith and George C. Williams (along with Ernst Mayr) in 1999. Not bad for people working in a field with "modest intellectual and scientific content." It does make you wonder, though. If intelligent design theorists are really making breathtaking discoveries in a new paradigm, why isn't the Royal Swedish

Academy of Sciences beating a path to their doors to shower them with prizes? No doubt the Swedish Academy is in cahoots with the American National Academy of Sciences in the naturalistic conspiracy.

Reading Johnson you would never guess that there is a whole army of geneticists, behavioral ecologists, paleontologists, evolutionary biologists, conservationists, agricultural scientists, medical scientists, developmental biologists, molecular biologists, physiologists, psychologists, and many others doing scientific research every day using Darwinian principles. They apply for grants and publish in a multitude of refereed professional journals.

Compare this to the virtually nonexistent fruits of intelligent design theory. Why are there no books like: *Animal Behavior: An Intelligent Design Approach, The Intelligent Design of Sperm Competition, Infanticide in Nature: An Intelligent Design Perspective, The Intelligent Design of Parasites,* or *The Intelligent Design of Aging?* Could it be that the theory is utterly useless for investigating the real world? Johnson says that:

> Materialists tend to think the only alternative to materialism is some form of primitive superstition, where science would be impossible because all events would be produced by the whimsy of capricious gods. This is nonsense, of course. Intelligent design does not mean unintelligent chaos. Computers and space rockets are designed, but they work according to lawlike principles.

We can see who designed computers and space rockets, and both they and their designers obey the laws of physics and chemistry. Johnson says that it is "nonsense" to say that intelligent design would involve the "whimsy of capricious gods." But as we've seen, in the one case where he tries to explain anything, *he* appeals to a "whimsical Creator." More importantly, as we saw in a previous chapter, Johnson never tells us what his criteria are for when the supernatural is allowed in and when it's not. Did the Creator intervene only once, or multiple times? Does supernatural intervention take place on a regular basis? What is his evidence that it did or does?

Where are the published fruits of intelligent design theory? Are there any articles or books which explain *anything* in the natural world? What are the contributions of intelligent design to the study of animal behavior? How does it explain sibling rivalry or mate guarding? What predictions would it make concerning when diseases would be most virulent?

Near the end of *Defeating Darwinism* Johnson predicts that evolutionary theory will ultimately "collapse with astonishing swiftness." Just what will cause this demise?

> The beginning of the end will come when Darwinists are forced to face this one simple question: *What should we do if empirical evidence and materialist philosophy are going in different directions?* (emphasis in the original)

He then informs us that, "of course," they are. And what is this astonishing evidence that refutes materialist philosophy? You've guessed it. There isn't any. Johnson hasn't produced *any* evidence that refutes materialism or evolution or that demonstrates supernatural intervention in nature. What he has produced is a long litany of negative argumentation mostly involving the Arguments From Personal Incredulity and Ignorance. He hasn't produced one empirical fact that refutes Darwinism. He hasn't produced even one positive argument for intelligent design.

While we're on the subject of important questions, we might make a short list of questions for advocates of intelligent design. What is a "type" or "kind?" What specific mechanisms limit the amount of change in a "type" or "kind?" What specific mechanisms cause mutation and natural selection to stop so that "types" are preserved (whatever they are)? How much change does there have to be before microevolution becomes macroevolution? Do you have any evidence of supernatural intervention in nature? Do you have any evidence of organisms popping into existence without ancestors? Why does sex exist? Why do males in many species kill infants sired by other males? Why do males in many species live shorter, more violent lives than females? Why do animals give preferential treatment to relatives? Why are the majority of fertilized eggs spontaneously aborted? Why do some fish switch sexes during their lifetimes? Why do humans have larger testicles than gorillas, but smaller testicles than chimpanzees? Why are so many sperm necessary? Why do forest trees grow so tall? Why do fish school, birds flock, and zebras herd? Why do insects have such intricate genitalia? Why do the canines in your mouth have such long roots? Why is human childbirth so painful? Why are humans so susceptible to choking?

I could go on for quite some time but the point I want to make is that evolutionary biology can explain, or at least suggest lines of research which might find an explanation, for all these phenomena. Some of the explanations are extremely well supported by empirical evidence. Others are more speculative and more research is needed. Evolutionary biology is a real science which makes real, testable predictions. On the other hand, if you pose any of these questions (or any other questions about the natural world) to an intelligent design theorist, I suspect that you will get no answer at all (except for: because the intelligent designer made it that way). Johnson (except for the lame example above) never gives us his alternative to evolutionary explanations.

I do want to cover one phenomenon in some detail because it perfectly captures the utility of evolutionary theory and the bankruptcy of intelligent design. The thing I want to discuss is aging (referring not just to the passage of time, but the deterioration of our bodies with time—the technical term is senescence). It is a process that, unfortunately, we all go through. In fact, it seems that all organisms that reproduce sexually gradually deteriorate over time and die (organisms that reproduce by splitting apparently do not age). Why do we grow old and eventually die? I am referring here to the "ultimate" reason, not the "proximate" reason(s). Ultimate questions are "why" questions. Proximate questions are "how" questions. There are many "how" theories on aging (free radicals, etc.), but we're interested in the "why" of aging.

As you might expect, there is (as far as I know) no intelligent design explanation for aging. I presume that no one knows why the designer wants his creations to waste away and die. I suspect that most intelligent design advocates would attribute aging and death to the sin of Adam in the Garden of Eden, although they would never state so explicitly (at least not when they are proposing intelligent design as a scientific theory). Presumably other animals were dragged down with us. Needless to say, this theory of aging has no evidence to support it and is not terribly useful in gerontological research.

Another theory is that we simply wear out with time like cars or washing machines. But this can't be right. For the first part of our lives, we actually get more robust with age. Furthermore, our bodies are quite capable of healing and regenerating themselves. New materials are constantly cycling through the body so that most of the atoms currently in your body have been there for a fairly short time. In addition, we are all descended from lines of cells that are hundreds of millions of years old. Sea anemones which reproduce by fission do not deteriorate with age. It is clearly possible to produce biological systems which do not deteriorate over time. Our mechanisms for maintenance and repair are what break down as we grow older. It would seem that maintaining our bodies would be a lot easier than building them in the first place. Mysteriously, whatever it is that creates the bodily adaptations that keep us alive seems incapable of maintaining them once they are in place.

One kind of evolutionary theory that you may have heard of is the idea that, for the good of the species, old individuals need to die off to avoid competing for resources with younger individuals. Biologists do not take this theory seriously because it involves what is called a "group selection fallacy." As we've already discussed, things do not evolve for the good of the group or species. To understand why, think of individuals in a population who happen to have genes which do not cause them to age and die.

In reproductive competition with individuals who do have genes for these effects, it is obvious whose descendants will inherit the earth.

So how does evolutionary biology explain aging? The answer was worked out by Peter Medawar, George C. Williams, and W.D. Hamilton. Consider a hypothetical population of animals which do not deteriorate with age. They are capable of reproduction at age 5 and the only way these animals ever die is by accident, disease, predation, etc. The inevitable result of these ever present hazards is that the number of individuals surviving to age 5 is higher than the number which survive to age 10 which is higher than the number which reach 15 and so on.

Now consider two gene mutations, one which causes death at age 2 and one which causes death at age 20. The gene which causes death at age 2 will be eliminated by natural selection since any individual with this gene will die before reproducing. The gene which is lethal at age 20 will be very weakly selected against and may not be selected against at all. The reason is that any individual who has reached age 20 has already reproduced and passed on the gene. The individual loses all chance of future reproduction, but the effect on the frequency of the gene could be trivial because most or all members of the population have died of other causes before they reach the age of 20.

Many genes have multiple effects, a phenomenon known as pleiotropy. Using our same hypothetical scenario, think about another genetic mutation. It causes increased fertility from age 5 onward but results in death at age 15. How will it do in competition against a gene which results in a constant fertility from age 5 onward? The answer depends on how much fertility is increased and how likely it is that individuals survive to age 15, but if survivorship is low enough the pleiotropic gene will be favored by selection even though it causes harmful effects at a later age. On the other hand, a pleiotropic gene which causes beneficial effects later in life, but detrimental effects early in life, will be selected against.

In all these examples, the detrimental effects need not be as dramatic as death. They could be a slight decrease in overall vigor. The key point is this: once the age of reproductive maturity is reached, the force of natural selection inevitably declines with age. This decline in the strength of natural selection results in the accumulation of late-acting detrimental mutations as well as the buildup of pleiotropic genes with beneficial early effects but detrimental later effects. Population geneticists have worked out the mathematics and it turns out that aging inevitably evolves in any sexually reproducing population. Furthermore, the rates of decline in various physiological systems will on average be very similar because genes which lead to more rapid deterioration in one system—as opposed to genes associated with slower declines in other systems—will be more strongly selected against. Selection will thus slow the rate of decline in the affected system until it is more in line with the rates of other systems. Evolution is inherently biased toward the young at the expense of the old. All of our ancestors were once young, but most of them were never what we would consider old.

This is all great in theory, but can it be tested against the real world? Yes, the theory can be, and has been tested in multiple ways. One way it has been tested is in selection experiments in the lab. If the theory is correct, you should be able to evolve longer-lived populations by progressively selecting for delayed reproduction. If you force a population to reproduce at later and later ages you force selection to maintain vigor and fertility at these later ages. Experiments like these have confirmed the theory by producing fruit flies with double the normal life span.

Another way to test the theory is to compare populations of the same species, one which is exposed to predators and one which is not (on an island, for example). After many generations have passed the population not exposed to predators should age more slowly since the number of individuals surviving to advanced ages will be substantially increased and the force of natural selection at these advanced ages will thus be increased. Steven Austad observed this exact effect in opossums.

The theory has also been tested using the comparative method. Animals with unusually effective defenses against predators should live longer than comparably sized animals who are more susceptible to predators. They do. Birds and bats deteriorate slower (they can escape through the air) than comparably sized non-flying mammals with comparable metabolic rates. Porcupines deteriorate slower than comparably sized mammals. Turtles are exceptionally long-lived.

The theory has been tested in other ways, and has important implications for how research into retarding aging should be done (see the notes for some references). My point in elaborating on aging is to highlight the differences between intelligent design theory and modern evolutionary theory. Here is a phenomenon of immense importance to all of us. It is utterly inexplicable using intelligent design. Evolutionary theory, on the other hand, can explain it. The evolutionary theory of aging has been tested and confirmed. It also has important implications for gerontology. Once you have evolved a longer-lived population of animals in the lab, you can then try to find out what it is about their biochemistry that enables them to deteriorate at a slower rate than the original population.

Finally, I mentioned earlier that it seemed strange that the force which "designed" the wonderful adaptations which keep us alive is capable of the prodigious feat of developing these adaptations from a single cell but is incapable of the seemingly much easier task of maintaining them once they are in place. We now know that that force is natural selection. When that force is weakened, as is inevitable with age, our adaptations progressively deteriorate until we are unable to sustain life even when completely at rest and supplied with oxygen and all necessary nutrients. We are not programmed to die at a certain age. We inevitably die when our evolved adaptations have deteriorated to the point where even the slightest insult is more than the body can handle. *Apparently, natural selection is the only thing which creates and maintains the adaptations which keep us alive.*

Hopefully, if you are curious about the other questions I mentioned earlier, or any other questions in biology, you will go down to a museum, library, or bookstore and find out for yourself. I'm all for the open minds which Johnson advocates. Johnson likes to quote Dawkins as follows: "It is absolutely safe to say that, if you meet somebody who claims not to believe in evolution, that person is ignorant, stupid or insane (or wicked, but I'd rather not consider that)." What Johnson fails to tell you is that immediately after that statement, Dawkins had this to say:

> If that gives you offense, I'm sorry. You are probably not stupid, insane or wicked; and ignorance is no crime in a country with strong local traditions of interference in the freedom of biology educators to teach the central theorem of their subject. I recently toured East Coast radio stations, doing phone-ins. I came away optimistic. I had expected hostile barracking from creationists with closed minds. Instead, what I found was genuine curiosity and honest interest. I got sincere questions from intelligent people who really wanted to know because they had literally no education in evolution.

I think most of the people who reject evolution do so because they haven't been exposed to it, or because it conflicts—or at least they think it conflicts—with their religious beliefs, a topic for the next chapter. Johnson, I believe, falls into this last category. At any rate, I would hope that you would not take Johnson's word or my word for it. Please do the research yourself.

We have seen what "God's Utility Function" is. Even if there were no fossil record, no molecular evidence, etc. we could still look at the real world to see if it looks like a world that was produced by evolution by natural selection. The evidence from animal behavior and other phenomena clearly indicates that it does. But what about the chapter title? What is Johnson's Utility Function? What is he maximizing in his attacks on evolution? It is the same thing that every lawyer maximizes when he has no case: obfuscation.

The Decline of Western Civilization

Unfortunately, in any discussion of creation/evolution you must finally get around to the religious question, since despite protestations to the contrary, this is virtually always the reason for creationist attacks on evolution. Johnson is no exception. In *Defeating Darwinism* he discusses the evils supposedly brought about by "modernism," the "established religion of the West." My chapter title is in jest, but Johnson clearly believes that Darwinism is the linchpin in a series of factors (the others being *Inherit the Wind* and the Supreme Court's school prayer decision in the 60's) which have led to all sorts of nastiness. Of course, people did bad things long before Charles Darwin or the 60's came on the scene. Nevertheless, there are large numbers of people who believe that the teaching of evolution leads to immorality. They usually say something like this: "If evolution is true, there is no God. If there is no God, there is no basis for morality. If there is no supernatural basis for morality, then why shouldn't everyone rape, murder, steal, cheat on their spouses, cheat their business partners, etc.?" The people who make this argument are basically saying that if there were no God to offer reward or threaten punishment they would start doing all sorts of nasty things. In other words, the only reason they behave in a decent manner is because they anticipate a supernatural reward or fear supernatural punishment, or both. No doubt many of these people would object to my characterization, and insist that they would be moral even if there were no supernatural (and thereby torpedo their own argument that there must be a supernatural basis for morality). Fair enough. But at the very least, they are accusing other people of being moral degenerates who must be threatened with a supernatural stick or paid a supernatural reward to keep them in line. I call this the Moral Prostitution Argument. This attitude was perfectly captured by a commentator on Pat Robertson's *700 Club* who, after reporting on the discovery of *Australopithecus garhi* (a hominid fossil), insinuated that we should expect things like the shootings at Columbine High School in Littleton, Colorado, if evolution is taught in the schools. Congressman Tom DeLay voiced a similar sentiment.

I find the people who make this argument to be somewhat frightening. If they are to be taken at their word, I certainly don't want to be around them if they ever lose their faith. Let's examine their claims in two stages. First, does evolution mean there is no God? No, as Carl Sagan, Ann Druyan and others have noted, evolution is consistent with atheism, but it doesn't imply it. Chemistry, physics, astronomy, auto mechanics, psychology, geology, medicine, rocketry, and a bunch of other things you can think of are all consistent with atheism, but they don't imply it. Revealingly, evolution is singled out for abuse. The reason is that evolution is in conflict with some particular versions of Christianity (including Johnson's). (I should note that I mean no offense by omitting discussion of all the other religions and creation stories. There are obviously many different versions of creationism. The dispute over the remains of Kennewick Man nicely illustrates the point. Kennewick Man is a 9,000-year-old skeleton discovered in Washington State. A dispute erupted when scientists wanted to study the skeleton, but local Indian tribes wanted the bones immediately reburied. The Indians believe that the remains are from one of their ancestors, and they don't like the idea of scientists studying the bones because they believe that their ancestors have been in America since the time of creation. I suspect that Johnson would dismiss this version of creationism as a myth, but if he did, I would be curious how he would draw the distinction between this version and his version. He has presented no more evidence for his version than the Indians have for theirs.)

Johnson wants Christians to buy his belief that if evolution is true, then Christianity is out the window. Given that there are Christian evolutionary biologists, as well as vast numbers of other Christians who accept the truth of evolution (including the Pope, who accepts the evolution of the human body, but adds that a soul was injected at some point), this belief is obviously false. One thing which contributes to this false belief is the so-called "Darwin Fish" emblems. I know they're intended to tease fundamentalists who attack evolution, but by mocking the Christian fish symbol they give the impression that you can't accept evolution and still be a Christian. The bottom line is this. Yes, there is a conflict between evolution and religion if your particular religion *requires* that humans and all other organisms not be related by common ancestry and not be the result of a long line of reproduction with modification. Johnson's religious beliefs require that descent with modification and common ancestry be false. Many other religious beliefs do not.

Johnson would object here and claim that I am evading the issue because what scientists mean (according to Johnson) when they talk about evolution is a process which absolutely excludes God from any involvement. As he says, "They are absolutely insistent that evolution is an *unguided* and mindless process, and that our existence is therefore a fluke rather than a planned outcome (emphasis in the original)." He cites a statement by the National Association of Biology Teachers which said that evolution was an "unsupervised, impersonal, unpredictable and natural process" (subsequently, due to theological pressure, the words "unsupervised" and "impersonal" were eliminated). Strictly speaking, he has a point here. But it is a fairly trivial point. What the scientists should say (if the subject comes up at all) is that there is no *evidence* that evolution is guided, supervised, or personal.

Consider the following statement: "Table salt crystals separate into sodium and chloride ions when dissolved in water. This is an unsupervised, impersonal, and natural process." Or, "Your son has died because his heart stopped beating due to blockages in his coronary arteries which deprived the cardiac muscle of oxygen, thus causing the tissue to die and rendering the heart incapable of contracting efficiently. This was an unsupervised, impersonal, and natural process." Strictly speaking, you cannot rule out supernatural influence in the previous two examples, and theologians should immediately insist on removing the second sentence in each case. Somehow I doubt whether anyone would object to using the phrase in these statements.

All we can say in these examples and in the evolution example is that there is no evidence of supernatural involvement and that injecting the supernatural would be superfluous. The living world and its history can be explained by an "unsupervised, impersonal, unpredictable and natural process"and there is no evidence of a "planned outcome," but that doesn't mean there is no God or that supernatural involvement is necessarily ruled out, only that it is unnecessary to explain the phenomena. Science cannot tell you whether God exists or not, nor can it tell you whether He is actively involved in nature or not. If you think it can, you're looking in the wrong place.

We have covered the "evolution means there is no God" claim. What about the Moral Prostitution Argument, the idea that there must be a supernatural basis for morality or else people will not behave themselves? If you think this is a good argument, ask yourself this question: if there were no supernatural basis for morality, would I immediately alter my behavior and start harming people by doing things which I previously wouldn't have done because I considered them to be immoral? In my opinion, if your answer to this question is yes, you deserve contempt (and are confirming that the Moral Prostitution Argument is aptly named). On the other hand, if your answer is no (and I think and hope that for almost everyone it would be), then you are implicitly stating that there does not have to be a supernatural basis for morality.

If you must stick with a reward and punishment system of morality remember that even if people wanted to rape, pillage, etc. they can be, and always have been, opposed by other people whose interests they threaten. You don't need supernatural entities for moral systems with methods of enforcement to develop. Cooperation has evolved not only in humans, but in many animal species. Humans are often singled out as particularly nasty compared to other animals, but humans in modern societies are not nearly as murderous as most species. Societies do have a beneficial effect. Ironically, it is often the very people who so adamantly oppose evolution who want to set up society in the most cutthroat, Darwinian way. Remember, evolutionary biology explains the world as it is. It is not a prescription for how society should be set up. I certainly don't want to live in a society set up the way nature is set up. We are fortunate to live in a society which at least partly insulates us from the real world. Surely, as Thomas Huxley put it, society should be directed toward "…not so much to the survival of the fittest, as to the fitting of as many as possible to survive."

If you're still troubled by the supposed evil effects of evolution, try a simple thought experiment. Imagine you have whatever religious beliefs (or lack thereof) you currently have and that evolution never happened. *Poof!* Now evolution did happen (and is happening). I'll bet you still love your family. The world continues on just like it did a minute ago. I'll bet you don't have a sudden urge to rape or kill. If you were an honorable person a minute ago, I'll bet you still are. If you were a jerk a minute ago, you still are. Society won't collapse because humans share a common ancestor with other organisms.

The single most important tenet of the Christian faith is the Resurrection. Did Jesus actually rise from the dead or did he not? Evolution has absolutely no effect on whether that historical event occurred or not. It also can't tell you whether humans have souls or not, or whether there is life after death.

Besides the religious issue, there is one other thing concerning evolution which seems to really upset people. I touched on it briefly earlier. It is the fact that humans descended from apes (not currently living apes, of course, but a common ancestral ape shared with chimpanzees). Why people are so troubled by this is baffling to me. I remember seeing *Planet of the Apes* and *2001: A Space Odyssey* when I was eight years old and being utterly fascinated. (Actually, the australopithecines in *2001* were a little scary at the time. The crescendo at the appearance of the monolith didn't help.) I have been interested in primates in general and apes in particular ever since. I think most kids are. It is the parents like the man (mentioned in an earlier chapter) who complained about the Tulsa zoo's chimpanzee exhibit who usually drum this wonderment out of them. What a shame.

I think most people would be enthralled if they took the time to learn about the great apes. They are remarkably like us, sharing many of our qualities, both good and bad. They recognize themselves in mirrors. Chimpanzees regularly engage in politics as they form coalitions and maneuver for social status. They have been known to die from grief. They cooperatively hunt and kill monkeys. They engage in warfare against adjoining groups. They make and use tools for various purposes. They search out and eat plants with medicinal qualities. They cooperate with and comfort each other. They deceive each other. They have been seen to make heroic efforts to save each other. They engage in rambunctious play and have a behavior which looks very much like human laughter. They plan out future courses of action. Different chimpanzee groups have different cultures (different types of tools, etc.) which they pass on to the next generation.

The other great apes (bonobos, gorillas, and orangutans) have their own interesting characteristics. I think if you make the effort to find out about them, you will find it an enjoyable experience.

How to Sink a Dinghy

Let's suppose that you're a clever lawyer. You have deeply held religious beliefs. Your particular version of Christianity requires that humans have been supernaturally separately created from all other creatures for a divine purpose. You believe that humans are the culmination of creation. You don't rule out the possibility that the Genesis chronology is true. You believe that if evolution is true, then Christianity is out the window. You also believe that the acceptance of evolution is responsible for many of the ills of society. You believe that humans are fundamentally different from animals and are not a part of the animal world. You reached this conclusion before you really started examining Darwinism. You have taken a hard look at the evidence for evolution, and decided that the evidence for it is weak if not existent. You have concluded that the only reason that scientists accept evolution is because they have a naturalistic bias. You believe that anyone who thinks that Darwinian evolution is true is engaged in self-deception. They are engaged in idolatry. They are trying to get rid of God, exactly as described in Romans, chapter one. How do you win the case against this evil philosophy? How do you sink the "great battleship" that is Darwinism?

First, there's jury selection. You know you'll never convince the scientific community. They look at you like you're nuts. They wonder if you really believe what you're saying. This naturalism is terribly blinding. It would be much better to make your case directly to the public, which is—fortunately for you—generally ignorant of evolutionary biology. Even better, many of them have religious beliefs similar to yours and do not want to believe that evolution is true. So a good strategy would be to publish popular books and articles and avoid like the plague publishing detailed expositions of your intelligent design theory of creation in the technical scientific literature.

Second, what kind of case should you put on? Should it be positive, negative, or both? Well, your only positive source of data on the matter is Biblical revelation. You could cite Romans and the Gospel of John, but then the game is over as far as appearing to be an unbiased analyzer of the evidence. Furthermore, bringing the Bible into the matter is bad politically, since it torpedoes your efforts to get your theory into public school science classes. The answer is easy: stick to negative arguments.

Now, what do you need to attack? You've got to show that the close biochemical similarity of all organisms and the branching pattern displayed by that similarity does not support evolution. The problem is that the idea that all organisms have descended from a common ancestor perfectly explains the data, and requires only that you assume that organisms do not pop into existence from nothing. Neither you nor anyone else has ever seen any living thing appear from nowhere so you can't attack this fact. What do you do? Turn things around. Say that scientists are merely assuming that ancestors existed because of their naturalistic bias and their prior commitment to Darwinism. Maybe your readers won't realize that scientists aren't assuming ancestors existed. They are looking at the real world, assuming that the same "laws of nature" applied in the past, and *concluding* that the connecting ancestors existed.

You've also got to show that the fossil evidence doesn't support Darwinian evolution. The fossils definitely show forms that don't exist today. No help there. Current organisms disappear from the record as you go back in time. Once again, no help. Fossils definitely show changes through time. They definitely show a logical progression through time. You don't have an out-of-place fossil. That doesn't help either. The record shows some transitions that even you have to admit are plausible. What do you do?

Well, there's that old reliable: gaps in the record. Just harp on the fact that there had to be scads of transitional forms in the past and try to get the jury to believe that most or all of them should have been preserved in the fossil record. Emphasize gaps created by Linnaean classification. It's either a reptile or a bird—with a gap in between. It's either an ape or a human. Strive hard to imply without stating it explicitly—for you might appear silly—that where gaps exist, the organisms involved mysteriously had no ancestors. Just hope that your readers don't realize that biology is full of continua and that no non-arbitrary lines can be drawn in a series of parent/offspring connections. Maybe your audience really will believe that organisms can have no ancestors even if that's never been observed in the real world. The beauty of the gaps strategy is that you can never lose. When one of your favorite gaps gets filled (whales, for instance), there are still—and forever will be—many more. Just hope that your readers don't realize that a complete, continuous, gapless fossil record would require a continuous sedimentary rock record several hundred miles thick. Above all else, hope they don't wonder why there should be any transitional forms at all if evolution is not true.

Next, you've got to attack the mechanism of evolution. Unfortunately, there's a problem here. Natural selection has been abundantly documented in the lab and in the wild. It has produced rates of evolution much faster than are necessary to explain the observed rates of change in the fossil record. It has proven enormously successful and useful in the study of animal behavior. You can't claim that mutation and natural selection don't happen. Even you don't believe that. What do you do? Draw a distinction between micro- and macroevolution! Imply that there is a huge chasm, an order of magnitude difference between the two. Imply that one has nothing to do with the other. You'll admit that mutation and natural selection can account for microevolution. Anyone can see that. But now comes the master stroke, the strategy which is as unbeatable as gaps in the record was for fossils. Effectively, you will define microevolution as any evolution that scientists have directly observed and macroevolution as any evolution they haven't observed. Now you can say that there is no mechanism for macroevolution since no one has ever experimentally shown that mutation and natural selection can account for macroevolution. It's foolproof. No matter what science comes up with, you just call it microevolution. Macroevolution remains forever mysterious and unexplained. Just hope that your readers don't realize that your distinction between micro and macro is totally arbitrary. Maybe they won't know that ultimately the differences between organisms are the result of differences in the "letters" in their DNA, and that there is no way to draw a line between the number of letter changes that are "micro" and the number required to be "macro." Hope they don't see that you have provided no specific mechanism for limiting change. Hope they don't see that you have no mechanism for causing mutation or natural selection to stop, something that has never been observed. And last, hope they don't realize that any historical science has to reconstruct the past by using mechanisms observed in the present. Maybe they won't catch on that what you are essentially saying is that no one should believe evolution or plate tectonics until scientists have evolved a mammal from a single-celled organism in the lab or directly observed the building of a mountain range in a lab.

Now that you've destroyed the hapless evolutionist case, you need to supply the jury with a reason why these deluded biologists believe such a silly thing as evolution. After all, it does seem strange that there could be such a monolithic conspiracy among scientists. Scientists are just like other people. They are ambitious and would like nothing better than to make earth-shattering, paradigm-shifting discoveries. They love to win Nobel and other prizes. Why can't they see the obvious fact that Darwinism is a house of cards? Of course! They have a naturalistic bias. If they didn't arbitrarily rule out the supernatural, they would immediately see that creationism provides a superior explanation for the history of life. Just hope that your readers don't realize that many scientists have religious beliefs. Hope they don't remember that invoking the supernatural has never explained anything in science and has been an impediment to progress in every case. Hope they don't realize it, but everyone has a "naturalistic bias" in everyday life. When their car breaks down, they don't curse the gods. They go to a mechanic. When they get sick, they go to the doctor. When they have a house with a broken window and numerous items missing, they don't suspect ghosts or spirits are the culprits. Strangely, of all the sciences and other fields of ordinary human activity, it is only biology that requires the supernatural. Maybe your readers won't notice that. Most importantly, hopefully they won't notice that it might be your own supernaturalistic bias which causes you to attack evolution.

Finally, we come to the most important part of your case. Whatever you do, *never* give anything even close to a specific scenario explaining your theory of the history of life. Never give any specific example where your theory is a better explanation than evolution by natural selection for any biological phenomenon. In short, never give your alternative to evolution. Set up a false dichotomy which hopefully your readers won't catch. Constantly imply that it's either evolution by natural selection or your own vague intelligent design creationism. If evolution is wrong, then intelligent design creationism is proven by default. Hope they don't realize that even in your camp you have massive, irreconcilable differences (Behe accepts common ancestry; there are young and old earth creationists, etc.). Hope they don't realize that there are hundreds of different creation stories. Maybe they will forget about theistic evolutionists. Perhaps they won't realize that even if evolution by natural selection were not true, there could still be a natural explanation for life. Maybe they won't realize that creationism doesn't explain anything, doesn't predict anything, and has never led to the discovery of anything new about nature.

You've got to avoid at all costs having your readers actually think about what your alternative to evolution might be, and about how it would square with the facts of natural history and the real world. The problem is, if you actually give an alternative to evolution, if you actually apply your intelligent design theory to the facts of the real world, you might look monumentally silly. How did all those fossils get there? They are totally unrelated to each other. They were intelligently designed, supernaturally popped into existence, and then, mysteriously, they went extinct. The designer went back to the drawing board, redesigned, and popped a new species into existence from nothing which was just slightly different from the first. Then that species went extinct. This cycle goes on over and over, supplemented periodically by mass extinctions. This goes on for billions of years, until the designer creates humans, the culmination of his plan. Your readers might ask what took the designer so long. They might wonder why humans have to be so damned similar to those nasty apes. They might note that the most charitable thing they could say about your "mechanism" is that it's about as missing as they come. They might ask why this "mechanism," which had to act millions of times in the past, is mysteriously never seen to act in the present. They might ask why the designer had to make things so that the only way for animals to stay alive is to eat another living thing. They might wonder if he enjoys cruelty, since he has designed things so that the vast majority of his creatures suffer and die before reproducing. They might wonder why he has designed things so that the world looks exactly like it would look if it had been produced by evolution by natural selection. These kinds of annoying questions could go on forever. It would definitely be better to avoid subjecting your theory to any contact with the real world. Just sit back and take shots at the other side.

If I were on the jury, I would be insulted and more than a little irritated that my time had been completely wasted. Johnson likes to envision the sinking of the Darwinian battleship and the imminent demise of Darwinism. Evolutionary biology will sail on, however. On the other hand, Johnson's intelligent design dinghy sinks itself by supernaturally developing holes in its phantom hull.

A Final Plea

I have a request to make. In this book I have tried to very concisely disperse the obfuscatory smokescreen that Phillip Johnson and other "new creationists" have raised in the last few years. If you desire a more in-depth, scholarly treatment of these topics I strongly recommend the recent books which I mentioned in the introduction: Robert Pennock's *Tower of Babel: The Evidence Against the New Creationism* and Kenneth Miller's *Finding Darwin's God*. Here is my request. Please don't take my word, or Johnson's word, or anyone else's word concerning creation/evolution. *Go look for yourself!* I have no illusions of convincing diehard creationists, but if you are honestly interested or concerned about this issue, I'm sorry, but you're going to have to do some work. The evidence for evolution by natural selection is publicly available. You can discover it for yourself. I have included very little of it here because I was more interested in analyzing Johnson's arguments. Do the research and you will discover a fascinating science which can deepen your appreciation for the natural world. If you are curious about why animals behave the way they do, make an effort to learn some evolutionary biology. There is a vast amount of literature on evolution and I have included in the text and the notes a selection of some of my favorites. Perhaps some of them can get you started.

If you are a student, parent, teacher, or school board member, I hope that you will push to have evolution given the emphasis it should have in junior high and high school biology. There is, of course, a big controversy about whether creationism should be taught in biology classes when evolution is taught. There are two reasons—besides the fact that the Supreme Court said it was unconstitutional—why it shouldn't be taught in science classes. First, on the few occasions when creationism has proposed testable hypotheses (such as that the earth is only a few thousand years old), they have been proven false. Second, when creationism proposes no testable hypotheses (as in Johnson's version), it invariably reduces to religion (or at least to not so subtle references to a Designer) and does not belong in a science classroom. Besides, which creation story would you use? Your typical creationist would scream if any creation story other than Genesis was mentioned. Unfortunately, what apparently often happens is that while creationism is justifiably left out, evolution is given little or no emphasis. The current creationist strategy is to remove evolution as much as possible from the curriculum.

Giving evolution the emphasis it deserves would greatly improve science education and hopefully somewhat ameliorate the dismal state of scientific literacy. If nothing else, it would expose students to the exciting science that biology can be. May students everywhere be spared the tedium of taking biology classes that don't include the central unifying theory of biology.

About the Author

Philip Frymire has degrees in zoology and geology from the University of Oklahoma. He has worked as a petroleum geologist for 16 years. He lives in Tulsa, Oklahoma.

Notes

Introduction

The Dawkins quote is on p. xvi of the foreword to the Canto edition of John Maynard Smith's *The Theory of Evolution* (Cambridge: Cambridge University Press, 1993). The Williams quote is on p. 167 of his book, *The Pony Fish's Glow* (New York: HarperCollins, 1997). Martin Gardner's comments on Johnson's objections to Darwinism are in his column in the *Skeptical Inquirer* 21: 6, pp. 17-20 (November/December, 1997). Johnson refers to a conference on "Mere Creation" on p. 93 of *Defeating Darwinism* (Downers Grove, IL: InterVarsity Press, 1997). Good refutations of creationism include Douglas J. Futuyma, *Science on Trial: The Case for Evolution* (New York: Pantheon Books, 1983) (a new edition of this book came out in 1995); Niles Eldredge, *The Monkey Business: A Scientist Looks at Creationism* (New York: Washington Square Press, 1982); Tim Berra, *Evolution and the Myth of Creationism* (Stanford, CA: Stanford University Press, 1990); Arthur N. Strahler, *Science and Earth History* (Buffalo, N.Y.: Prometheus Books, 1987) (a revised edition came out in 1999); and, as already mentioned in the introduction, Robert T. Pennock, *Tower of Babel: The Evidence Against the New Creationism* (Cambridge, Mass.: The MIT Press, 1999) and Kenneth R. Miller, *Finding Darwin's God* (New York: HarperCollins, 1999). Good websites in this vein include Talk.Origins at http://www.talkorigins.org, the National Center for Science Education at http://www.natcenscied.org, and "The World of Richard Dawkins" at http://www.world-of-dawkins.com.

The Materialist Conspiracy

Johnson compares the National Academy of Sciences to the College of Cardinals in an interview available on the web at http://www.arn.org/docs/johnson/commsp99.htm. On methodological and ontological naturalism, see *Tower of Babel*. Johnson's quote concerning "universal common ancestry" is on p. 94 of *Defeating Darwinism*, and his quote concerning "a God who acted openly" is on p. 23 of the same book. The uselessness of invoking God as a scientific explanation is covered in detail in Richard Dawkins's *The Blind Watchmaker* (New York: W.W. Norton, 1986) and his *Climbing Mount Improbable* (New York: W.W. Norton, 1996). Pennock's *Tower of Babel* goes into much more detail concerning the problems caused by letting the supernatural into science and other areas. The "materialism applied to the mind" quote is on p. 82 of *Defeating Darwinism*. The Richard Feynman quotation is on p. 184 of I.S. Shklovskii and Carl Sagan's book, *Intelligent Life in the Universe* (New York: Dell Publishing, 1966). Richard Feynman's views on materialism and purpose in the universe can be found in James Gleick's book, *Genius: The Life and Science of Richard Feynman* (New York: Pantheon Books, 1992) and in Richard Feynman, *The Meaning of it All: Thoughts of a Citizen Scientist* (Reading, Mass.: Addison-Wesley, 1998). The "Johnson's preaching" quote is from *Tower of Babel*, p. 300. The James Randi Educational Foundation (whose website is at http://www.randi.org) has a $1,000,000 reward for anyone who can demonstrate a paranormal phenomenon under controlled conditions. So far, no one has collected even though psychics and faith healers would have to do considerably less fund raising if they simply did what they claim they can do.

A Micromountain From a Macromolehill

For detailed evidence concerning evolution and natural selection see Mark Ridley's textbook, *Evolution* (second edition) (Cambridge, Mass.: Blackwell Science, 1996). The micro- and macroevolution quotes are from *Defeating Darwinism*, p. 57. The "Creationist farmer" quote is on p. 255 of Jonathan Weiner's book, *The Beak of the Finch* (New York: Alfred A. Knopf, 1994). The high praise this book received (including the Pulitzer Prize) seems to have really irritated Johnson since he uses the phrase "finch-beak variation" obsessively in various books and articles. Carl Zimmer's book on macroevolution is *At the Water's Edge: Macroevolution and the Transformation of Life* (New York: The Free Press, 1998). This book is also a good source concerning the relationship of development and evolution. For information on the "major transitions in evolution" see John Maynard Smith and Eörs Szathmáry's, *The Origins of Life* (New York: Oxford University Press, 1999). The "what sort of proof is this" quote is from p. 69 of Phillip E. Johnson's *Darwin on Trial* (second edition) (Downers Grove, IL: InterVarsity Press, 1993). See *Finding Darwin's God* concerning rates of evolution and the supposed "missing mechanism" for macroevolution. It is amusing to see Johnson complain about a missing mechanism when his "mechanism," whatever it is, has never been observed. For an excellent discussion of how to view ancestors/descendants and evolutionary trees see chapter 11 in Steven Pinker's *The Language Instinct* (New York: William Morrow, 1994). Besides, any book which includes phylogenetic trees containing Magilla Gorilla, Fred Flintstone, and Bonzo deserves to be read. Tijs Goldschmidt's book is *Darwin's Dreampond: Drama in Lake Victoria* (Cambridge, Mass.: The MIT Press, 1996).

The Fossil "Problem"

The "rarely inform the public" quote is from p. 38 of *Defeating Darwinism* and the "appeared mysteriously" quote is from p. 67 of the same book. Gould's book is *Wonderful Life* (New York: W.W. Norton, 1989). For critiques of Gould's views, see Daniel Dennett's *Darwin's Dangerous Idea* (New York: Simon & Schuster, 1995), Richard Dawkins's *Unweaving the Rainbow* (New York: Houghton Mifflin, 1998), and Simon Conway

Morris's *The Crucible of Creation* (Oxford: Oxford University Press, 1998). For Gould's views on punctuated equilibrium, see any of the excellent published collections of his *Natural History* magazine essays: *Ever Since Darwin* (1977), *The Panda's Thumb* (1980), *Hen's Teeth and Horse's Toes* (1983), *The Flamingo's Smile* (1985), *Bully for Brontosaurus* (1991), and *Eight Little Piggies* (1993), all published by W.W. Norton (New York). For critiques of punctuated equilibrium see *Darwin's Dangerous Idea, The Blind Watchmaker, Finding Darwin's God,* and John Maynard Smith's *Did Darwin Get it Right?* (New York: Chapman & Hall, 1989).

This is as good a place as any to address the controversy surrounding Gould's views. Johnson and other creationists love to quote Gould and emphasize his differences with Dawkins. In fact, Robert Wright refers to Gould as "The Accidental Creationist" in an article in *The New Yorker* (Dec. 13, 1999) in which he gives his view that Gould has unintentionally aided the creationist cause. That's going considerably overboard, I think. Here are several things to remember. First, there is no controversy within biology regarding descent with modification. Second, there is no controversy surrounding the fact that natural selection is the only known mechanism for producing complex adaptations. Third, although Dawkins and Gould are excellent writers, they do not represent the whole field, nor are their writings a substitute for the technical literature. Gould wrote an extremely negative review of Helena Cronin's *The Ant and the Peacock* (Cambridge: Cambridge University Press, 1991) in *The New York Review of Books* (Nov. 19, 1992) which drew replies from John Maynard Smith and Daniel Dennett. Johnson commented extensively on this review in his book, *Reason in the Balance* (Downers Grove, IL: InterVarsity Press, 1995) and used it to imply that there is a schism in biology. Having enjoyed Cronin's book myself, I was curious what professional evolutionary biologists aside from Gould thought of it. George Williams supplied a more interesting perspective in his review of Cronin's book in a professional journal, *The Quarterly Review of Biology* 68: 3, pp. 409-412 (Sep. 1993), entitled "Hard-Core Darwinism Since 1859."

Williams noted that he had already identified 19 published reviews of Cronin's book in various newspapers, magazines, and technical scientific journals. As he put it, "With one outstanding exception, the reviews are all favorable, some lavishly so." And later, "An obvious question arises: If the book is that much in error, why was Gould the only one who noticed?" Williams himself stated that he was "an ardent admirer" of the book. Gould published two more controversial articles in *The New York Review of Books*—you'll notice that the overheated rhetoric of both Gould and his critics is usually found in popular treatments, not technical journals—in June of 1997, arguing against "Darwinian Fundamentalism" and expounding on "Evolution: The Pleasures of Pluralism." These articles provoked several replies, but the most interesting one wasn't published. It was written by John Tooby and Leda Cosmides of the Center for Evolutionary Psychology at UCSB. It is on the web at http://cog-web.english.ucsb.edu/Debate/CEP_Gould.html. This passage captures their frustration: "For biologists, the central problem is that Gould's own exposition of evolutionary biology is so radically and extravagantly at variance with both the actual consensus state of the field and the plain meaning of the primary literature that there is no easy way to communicate the magnitude of the discrepancy in a way that could be believed by those who have not experienced the evidence for themselves."

I got the "evolution by jerks" joke from John Brockman's book, *The Third Culture* (New York: Simon & Schuster, 1995), p. 91. Steve Jones mentions "evolution by creeps" on p. 105 of his *The Language of Genes* (New York: Anchor Books, 1994). Stephen Jay Gould's book is *Dinosaur in a Haystack* (New York: Harmony Books, 1996).

Intelligent Design

The "Never say" quote is from Richard Dawkins's *River Out of Eden* (New York: HarperCollins, 1995), p. 70. The "fairly convincing" and "persuasive evidence" quotes are from Michael Behe's book, *Darwin's Black Box*

(New York: The Free Press, 1996), p. 5 and p. 15. The "molecular mechanisms" quote is from *Defeating Darwinism*, p. 77. Concerning the Argument from Ignorance, see *Tower of Babel* and a review of *Darwin's Black Box* entitled "Argumentum Ad Ignorantiam" in *The Quarterly Review of Biology* 72: 4, pp. 445-447 (Dec. 1997) written by Neil W. Blackstone. I encourage readers to think up their own examples of complex systems and how they could have evolved. Don't use only biological examples. Consider things like economies or cities. I'm sure you'll be able to come up with examples that at first seem totally inexplicable—especially to an observer seeing only the finished product—but that did evolve by gradual steps. The key, of course, is to keep thinking, rather than giving up. Behe's definition of irreducible complexity is on p. 39 of *Darwin's Black Box*: "...a single system composed of several well-matched, interacting parts that contribute to the basic function, wherein the removal of any one of the parts causes the system to effectively cease functioning." The "unintelligent material processes" quote is from p. 108 of *Reason in the Balance*. The quotes from George Williams concerning the evolution of the eye are from *The Pony Fish's Glow*, p. 13. The "suppose that nearly" quote (as well as the additional quotes concerning Behe's design scenario) is from *Darwin's Black Box*, pp. 227-228. The "more than a thousand," "preformed genetic information," and "what this colossal mistake" quotes are from *Finding Darwin's God*, pp. 162-163. On the evolution of mammalian ear bones see Stephen Jay Gould's essay, "An Earful of Jaw," in *Eight Little Piggies*, and Kenneth Miller's review of *Darwin's Black Box* available on the web at http://www.korrnet.org/reality/miller_review.html (originally published in *Creation/Evolution* 16: 2, pp.36-40). He also discusses it in *Finding Darwin's God*. One particularly good review included in "Behe's Empty Box" was written by H. Allen Orr in *The Boston Review* and is at http://www-polisci.mit.edu/bostonreview/BR21.6/orr.html. Johnson betrayed some of his ignorance of evolutionary biology in his reply to this review (also included in "Behe's Empty Box"). For a discussion of "information" which is far more interesting (and useful) than

Johnson's, see Steven Pinker's *How the Mind Works* (New York: W.W. Norton, 1997). This book also contains an explanation of natural selection, its implications, and how it could be falsified. The "just a trick" quote is from *Defeating Darwinism*, p. 74. The "meddling with the sex lives of dogs" quote is from Carl Sagan and Ann Druyan's book, *Shadows of Forgotten Ancestors* (New York: Random House, 1992), p. 54. The "learn to distinguish" and "if I say" quotes are from *Defeating Darwinism*, pp. 42-43. The quote from Martin Daly and Margo Wilson's *Homicide* (Hawthorne, N.Y.: Aldine de Gruyter, 1988) is on p. 3. This book is a real eye-opener showing how Darwinian theory accounts for who kills whom, and why. Sarah Hrdy discusses infanticide and many other interesting topics in her book, *Mother Nature* (New York: Pantheon Books, 1999). For examples of functionally stupid features that indicate organisms' history of trial and error tinkering, see *The Pony Fish's Glow* and Randolph M. Nesse and George C. Williams's book, *Why We Get Sick: The New Science of Darwinian Medicine* (New York: Times Books, 1994). The latter book emphasizes the importance and usefulness of Darwinian thinking in medicine. The giraffe's recurrent laryngeal nerve is discussed in Mark Ridley's *Evolution*. I got my information on the ductus arteriosus from Arndt von Hippel's *Human Evolutionary Biology* (Anchorage, AK: Stone Age Press, 1994). If you want a book on human anatomy and physiology that is fun to read, this is the book for you. The "inscrutable," "whimsical," and "mysterious" quotes are from *Darwin on Trial*, pp. 71, 31, and 67. The "multiple motives," "designers who have," and "reasons that a designer" quotes are from *Darwin's Black Box*, p. 223. For a comical look at "design", see E.T. Babinski's "Cretinism or Evilution?" at http://www.talkorigins.org/faqs/ce/4.

Human Evolution

Good general references on human evolution include Robert Boyd and Joan B. Silk's *How Humans Evolved* (New York: W.W. Norton, 1997) (revised edition 1999), Christopher Stringer and Robin McKie's *African Exodus* (New York: Henry Holt and Co., 1996), Alan Walker and Pat Shipman's *The Wisdom of the Bones* (New York: Alfred A. Knopf, 1996), and Christopher Wills's *Children of Prometheus* (Reading, Mass.: Perseus Books, 1998). As mentioned in the text, Donald Johanson and Blake Edgar's *From Lucy to Language* (New York: Simon & Schuster, 1996) contains beautiful life size photographs of many of the most important fossils. On molecular data in evolution see Roger Lewin's *Patterns in Evolution: The New Molecular View* (New York: W.H. Freeman and Company, 1999) and John C. Avise's *The Genetic Gods* (Cambridge, Mass.: Harvard University Press, 1998). The "share a common biochemical basis" quote is from *Darwin on Trial*, p. 91. The quotes from John Avise are on pp. 35 and 36 of *The Genetic Gods*. Michael Behe's acceptance of the common ancestry of humans and the great apes is described on p. 164 of *Finding Darwin's God*. As Kenneth Miller describes it, he was debating Behe and, "I presented him with molecular evidence indicating that humans and the great apes shared a recent, common ancestor, wondering how he would refute the obvious. Without skipping a beat, he pronounced the evidence to be convincing, and stated categorically that he had absolutely no problem with the common ancestry of humans and the great apes. Creationists around the room—who had viewed him as their new champion—were dismayed." Behe is critical of Darwinism, but unlike Johnson he does not deny the obvious. It would be interesting to know how many creationists—most consider Behe's book to be a godsend—are aware that Behe accepts that humans and the great apes share a common ancestor. Richard Dawkins's article, "Gaps in the Mind," is in *The Great Ape Project* (New York: St. Martin's Press, 1993) edited by Paola Cavalieri and Peter Singer. Dawkins also discusses "ring species" in this article. These are species which, in a local area, are clearly and obviously distinguishable. However, if you follow them around a geographical "ring" they become less and less like the species you started with and more and more like the other species in the pair. Dawkins uses the example of herring gulls and lesser black-backed gulls, which are distinct species in Britain. As you go west around the North Pole through North America, across from Alaska to Siberia, and back to Europe, the herring gulls become less and less like herring gulls and more and more like lesser black-backed gulls. The lesser black-backed gulls, which don't interbreed with herring gulls in Britain, are the other end of a ring which links the two species through a continuous series of interbreeding intermediates. Dawkins makes the key point: "The only thing that is special about ring species like these gulls is that the intermediates are still alive. *All* pairs of related species are potentially ring species. The intermediates must have lived once. It is just that in most cases they are now dead (emphasis in the original)." The "alleged hominid species" and "may also have been" quotes are on p. 85 of *Darwin on Trial*. The "If they could interbreed" quote is in a footnote on p. 61 of *Defeating Darwinism*. Christopher Wills discusses chimpanzee/bonobo hybrids in *Children of Prometheus*. The question of whether a human/chimp hybrid is possible is an interesting one. Humans have 46 chromosomes while chimpanzees have 48. Chromosome 2 in humans is a fusion of two ape chromosomes which results in the reduction from 48 (24 pairs) to 46 (23 pairs) chromosomes. Hybrids have occurred between species with different numbers of chromosomes and between species that are less closely related than chimpanzees and humans. Chimps and bonobos can interbreed and they last shared a common ancestor about three million years ago, or about half the amount of time that the human and chimp lines have been separate. I remember taking a primate behavior class in the early 1980's and being told that human/chimp hybrids might be "possible." I have checked several sources and if the topic is mentioned at all, it is usually stated that it is at least "possible" or "cannot be ruled out." Presumably the subject is such an ethical hot potato (would such a hybrid have "human" rights?) that no one has tried to bring sperm and egg together, or if someone has, he or she has not made the results known. There was a BBC miniseries (later shown in the U.S.) produced in 1989 entitled "First Born" which explored this topic. In the film a scientist produced a human/gorilla hybrid and faced a terrible moral dilemma. My memory of the film is vague, but I believe he debated whether to kill the infant hybrid but couldn't bring himself to do it. He subsequently tried to raise the child in human society (in the film the child looked externally human), but things went terribly wrong (as they always do in this genre). Similar ethical problems may come up once the Human Genome Project is finished. This is the effort to completely sequence the entire human genome. Once this is done, and if the complete chimpanzee genome is sequenced as well, it should be possible at some point in the future to determine exactly what genes account for the differences between humans and chimps. Scientists might then be tempted to start inserting human genes into chimps and observing the results. The "whole lot of evidence" quote is from *Defeating Darwinism*, p. 38.

Johnson's Utility Function

The quotes from *River Out of Eden* are on p. 105. The quote from *The Pony Fish's Glow* is on p. 154. George C. Williams covers the nastiness of nature in much more detail in his article "Huxley's Evolution and Ethics in Sociobiological Perspective" originally published in *Zygon* (1988) pp. 383-407, and reprinted in *Issues in Evolutionary Ethics* (Albany, N.Y.: SUNY Press, 1995), pp. 317-349, edited by Paul Thompson. This article is a condensed version of his essay "A Sociobiological Expansion of *Evolution and Ethics*" in *Evolution and Ethics: T.H. Huxley's Evolution and Ethics with New Essays on its Victorian and Sociobiological Context* (Princeton, N.J.: Princeton University Press, 1989). David Hull's review of *Darwin on Trial* was in *Nature* 352: 485-486 (Aug. 8, 1991). The "Darwinist doctrine of God" quote is from Johnson's reply to Stephen Jay Gould's scathing review of *Darwin on Trial*. Johnson's reply is entitled "The Religion of the Blind Watchmaker" and is available on the web at http://www.arn.org/docs/johnson/watchmkr.htm. Gould's review, entitled "Impeaching a Self-Appointed Judge," was in *Scientific American*

(July, 1992). The quotations from Johnson concerning peacocks and sexual selection are from *Darwin on Trial,* pp. 30-31. The quotation and data on better growth and survival in offspring of more elaborate peacocks is from a paper by Marion Petrie entitled "Improved growth and survival of offspring of peacocks with more elaborate trains" in *Nature* 371: 598-99 (Oct. 13, 1994). On "selfish genes" see Richard Dawkins's *The Selfish Gene* (new edition) (Oxford: Oxford University Press, 1989). George C. Williams's *Adaptation and Natural Selection* was reprinted with a new preface in 1996 (Princeton, N.J.: Princeton University Press). On the success of "selfish gene" theory, see Tooby and Cosmides's response to Gould cited previously, Matt Ridley's *Genome* (New York: HarperCollins, 1999), or any modern textbook on animal behavior. Good books on sexual selection include *The Ant and the Peacock,* Matt Ridley's *The Red Queen: Sex and the Evolution of Human Nature* (New York: Macmillan, 1994), and James L. Gould and Carol Grant Gould's *Sexual Selection: Mate Choice and Courtship in Nature* (New York: W.H. Freeman and Company, 1997). The "exist for the pure" quote is from *Genome,* p. 127. This book discusses in detail the battles (including those between the X and Y chromosomes) going on in the genome. Conflicts during pregnancy are discussed in *Genome* and *The Pony Fish's Glow,* as well as many other sources cited here. The "evolutionary biology is a field" and "materialists tend to think" quotes are from *Defeating Darwinism,* p. 113 and p. 96. To get a good idea of the importance of evolutionary biology consult the "White Paper" on evolutionary biology entitled "Evolution, Science and Society: Evolutionary Biology and the National Research Agenda" on the web at http://www.rci.rutgers.edu/~ecolevol/fulldoc.html. For information on the evolution of aging see *The Pony Fish's Glow, Why We Get Sick,* and Steven Austad's *Why We Age* (New York: John Wiley & Sons, 1997). If you want to get really technical, see Michael R. Rose's *Evolutionary Biology of Aging* (New York: Oxford University Press, 1991). This book covers the experiments with fruit flies in detail. The Dawkins quotes concerning people who don't believe in evolution are from *The New York Times* book review section dated April 9, 1989, p. 34 (he was reviewing *Blueprints*, by Donald Johanson and Maitland Edey). This review is also available on the web at http://www.world-of-dawkins.com/review_blueprint.htm.

The Decline of Western Civilization

Carl Sagan and Ann Druyan's comments concerning evolution and atheism are found on p. 66 of *Shadows of Forgotten Ancestors*. The "absolutely insistent" quote is on p. 15 of *Defeating Darwinism*. The National Association of Biology Teachers statement on evolution is on the web at http://www.nabt.org/Evolution.html. Good books on the evolution of cooperation include Robert Axelrod's *The Evolution of Cooperation* (New York: Perseus Books, 1984) and Matt Ridley's *The Origins of Virtue* (New York: Viking Penguin, 1997). The issue of whether morality comes from God is, of course, a very old philosophical problem. See *Tower of Babel* or any book on the philosophy of ethics for a more complete discussion. The question is whether things are morally good because God commands that they are, or does God command what is morally good? Creationists usually maintain the first option. But what if God commanded that we torture and kill small children? The reply usually given would be that God would never command such a thing. Why? Because it's not morally correct. But that means that the moral goodness of something exists independently of God's commands, which means that there is a basis for morality whether God exists or not. George C. Williams comments on the relative rates of intraspecific killing in humans and other animals in his article "Huxley's Evolution and Ethics in Sociobiological Perspective." The quote from Thomas Huxley is from an excerpt from his *Evolution and Ethics* reprinted in *Issues in Evolutionary Ethics*, p. 133. Interesting books on chimpanzees and other apes include Richard Wrangham and Dale Peterson's *Demonic Males* (New York: Houghton Mifflin, 1996), Frans de Waal and Frans Lanting's *Bonobo: The Forgotten Ape* (Berkeley, CA: University of California Press, 1997), and Roger Fouts's (with Stephen

Tukel Mills) *Next of Kin* (New York: William Morrow & Co., 1997). Fouts was a psychology professor at the University of Oklahoma when I was a freshman there. I volunteered to help out at the Institute for Primate Studies which he describes in his book. As Fouts details in his book, the conditions at the Institute were not the best. My first trip out there led to a memorable encounter with a large male chimp. I don't remember the chimp's name, but like most of the chimps at the facility he had been confined in less than ideal conditions for a long time. I was twice spat upon with full mouthfuls of water in between bouts of screaming and backhanded blows to an iron plate in his cage which sounded as if someone were hitting it with a skin-covered baseball bat. I was intimidated down to the marrow of my primate bones. Fortunately, feces did not fly. When you look into the eyes of a chimpanzee, there's someone looking back.

How to Sink a Dinghy

Johnson refers to Darwinian evolution as a "great battleship" on p. 169 of *Darwin on Trial*. Johnson's views on Christianity, evolution, and the self-deception of Darwinian evolutionists can be found in the previously cited interview available on the web at http://www.arn.org/docs/johnson/commsp99.htm. If you really want to see where Johnson is coming from, what really motivates him, and why he really attacks evolution, I urge you to read this interview. It becomes obvious that his only source for real "truth" is the Bible. He doesn't even rule out the Genesis chronology. To me, that speaks volumes. No one can seriously claim to be objectively evaluating the scientific evidence without religious bias if they refuse to simply state that the Genesis chronology is not literally true.